CHEMISTRY

An Industry-Based Laboratory Manual

T0346299

CHEMISTRY

An Industry-Based Laboratory Manual

John Kenkel

LEWIS PUBLISHERS

Boca Raton   New York   London   Tokyo

Published in 2000 by
CRC Press
Taylor & Francis Group
6000 Broken Sound Parkway NW, Suite 300
Boca Raton, FL 33487-2742

International Standard Book Number-10: 1-56670-346-8 (Softcover)
International Standard Book Number-13: 978-1-56670-346-8 (Softcover)
Library of Congress catalog number: 00-030171

Library of Congress Cataloging-in-Publication Data

Kenkel, John.
   Chemistry : an industry-based laboratory manual / John V. Kenkel.
     p. cm.
   Includes bibliographical references and index.
   ISBN 1-56670-346-8 (alk. paper)
   1.Chemistry—Laboratory manuals. I. Title.
QD45 .K34   2000
540—dc21   00-030171

Visit the Taylor & Francis Web site at
http://www.taylorandfrancis.com

and the CRC Press Web site at
http://www.crcpress.com

Preface

Have you ever sat down and had an open discussion over coffee with an industrial chemist or chemical technician about his or her work? Have you ever "shadowed" a chemist or a chemical technician for a day as he or she went about the daily routine in an industrial or government laboratory? If you have done these things, you were likely surprised at how foreign the language seemed or startled at how unfamiliar the surroundings were, despite what you perceived as your thorough academic grounding in the principles of chemistry. Is there any talk of the quantum mechanical model of the atom? No. Is there any activity relating to the molecular orbital theory of bonding? No. Is there any use made of your knowledge of Lewis acids and bases? Probably not. Is the lab a large room with six 12-ft lab benches capable of accommodating up to two dozen chemists? No, not usually. The labs are typically small, with from one to six lab workers in each.

Do you hear anything about safety in the laboratory. Yes! In fact, if you visited a lab, you probably had to go through some sort of safety check, such as reading basic safety guidelines or the viewing of a safety video, prior to even entering the facility. You probably also had to "sign in" and be issued personal protective equipment, such as safety glasses and a lab coat. If you later followed a chemist into the production facility to take samples, you may have been issued a hard hat or perhaps a small white disposable stretch-cap to cover your hair.

If you toured a lab facility, you probably toured a "wet lab," a "quality control lab," or perhaps a "process development lab," or maybe all of the above and wondered what these terms meant. You probably saw a control chart and wondered what it was. You may have sat in on a meeting to prepare for the upcoming quality assurance audit and wondered what an audit actually was or what GLP, MSDS, and SOP meant. You may have heard someone talk about certified reference materials and wondered what that was. You may have encountered a formal means of disposing of chemical waste and said, "Wow!" Or you may have noticed an experiment or an instrument that wasn't working properly and, subsequently, observed chemists and technicians teaming together for troubleshooting.

Yes, unless your course in chemistry was "industry-based," you probably came away from this experience with a thought that went something like this: "Wow, why don't textbooks and lab manuals do a better job of communicating what these real-world chemists do?"

The intent of the author of the lab manual you hold in your hands is to do just that. Students who progress through this lab manual will get a heavy dose of laboratory safety issues as they exist for the real-life chemist. Students who use this lab manual will keep records much like an industrial chemist keeps records and learn how important communication is to the chemist. Students who use this lab manual will learn what a control chart is, what certified reference material is, what is meant by GLP and MSDS, and what an SOP is all about. Students will read examples of what communications may occur between one industrial chemist and another and what is required to report the results of a laboratory analysis to a client. They will learn what an "out-of-spec" or "off-color" product is and how it may be studied. And, they will learn how to go about identifying unlabeled materials.

Students who use this lab manual will critically examine the labels on consumer products and commercial chemical products for safety information. They will attempt to solve problems for chemical companies and research proposed new industrial processes and laboratory methods. They will analyze materials and consumer products to provide third-party answers to industrial problems. They will participate in proficiency testing and identify a waste acid that is designated for disposal. Students also will learn how to think critically in order to apply chemistry principles to solve these various problems and to report to a client. Students even play the role of industrial chemists that become involved in National Chemistry Week activities.

Students who use this lab manual are laboratory technicians employed by a fictional consulting firm called the I.O.N.S. Corporation. The I.O.N.S. Corporation is directed by a chief executive officer by the name of Claire Hemistry, or C. Hemistry. The I.O.N.S. Corporation has a safety officer by the name of Ben Whell who issues a safety report for each project for which special safety considerations may be an issue. The instructor is the laboratory supervisor. Ms. Hemistry enters into contractual agreements with fictional clients that are faced with the problem or situation at hand. Industrial and academic consultants write memos giving the I.O.N.S. technicians the necessary background information. One of the consultants, the client, or Dr. Hemistry provides a standard operating procedure (SOP), or other procedure, that must be executed in order to solve the problem. The students then perform the lab work to solve the problem, keep a laboratory notebook according to company protocols, and write a report memo to the client on I.O.N.S. stationery reporting their results and recommendations, if appropriate.

Students do not know the outcome of the work before they start. In that sense, they are "discovery" experiments in which they do the work in order to "discover the answer" for the client. I hope both instructors and students will have as much fun working their way through this manual as I have had preparing it.

John Kenkel
Southeast Community College

Author

John Kenkel is a chemistry instructor at Southeast Community College in Lincoln, NE. He has a Master's Degree in chemistry from the University of Texas at Austin (1972) and a Bachelor's Degree in chemistry (1970) from Iowa State University.

Throughout his 23-year career at SCC, Kenkel has been directly involved in the education of chemistry-based laboratory technicians in a vocational program called Environmental Laboratory Technology. He also has been heavily involved in chemistry-based laboratory technician education on a national scale, having served on a number of American Chemical Society committees, including the Committee on Technician Activities and the Coordinating Committee for the Voluntary Industry Standards project. Besides these, he has served a 5-year term on the ACS Committee on Chemistry in the Two-Year College, the committee that organizes the Two-Year College Chemistry Consortium ($2YC_3$) conferences. He chaired this committee in 1996.

Kenkel has authored several popular textbooks for chemistry-based technician education. *Analytical Chemistry for Technicians* was first published in 1988. A second edition of this book was published in 1994. In addition, he has authored two other books, one entitled *Analytical Chemistry Refresher Manual* in 1992 and another entitled *A Primer on Quality in the Analytical Laboratory*, which was published in 2000. All were published through CRC Press/Lewis Publishers.

Kenkel has been the Principal Investigator for a series of curriculum development project grants funded by the National Science Foundation's Advanced Technological Education Program. A textbook (*Chemistry: An Industry-Based Introduction*) and this manual (*Chemistry: An Industry-Based Laboratory Manual*) were produced under these grants. He also has authored or coauthored four articles on the curriculum work in recent issues of the *Journal of Chemical Education* and has presented this work at more than a dozen conferences since 1994.

In 1996, Kenkel won the prestigious National Responsible Care Catalyst Award for excellence in chemistry teaching sponsored by the Chemical Manufacturer's Association.

Acknowledgments

Partial support for this work was provided by the National Science Foundation's (NSF) Advanced Technological Education program through Grant #DUE9751998. Partial support was also provided by the E.I. DuPont DeNemours Company through its Aid to Education Program. Any opinions, findings, and conclusions or recommendations expressed in this material are those of the author and do not necessarily reflect the views of the National Science Foundation or the DuPont Company.

The National Science Foundation support allowed a number of people to become involved in many different ways. The following attest to this.

Susan Rutledge, adjunct faculty member at Southeast Community College, very ably served as project assistant throughout. In this role, she researched and field tested experiments, developed and maintained a directory of chemical technology faculty members nationwide, created a valuable index for the American Chemical Society's (ACS) Voluntary Industry Standards document, and cheerfully accepted and accomplished many a menial task for the good of the project. She deserves much credit.

Don Mumm of Southeast Community College contributed two of the experiments and Karen Wosczyna-Birch of Tunxis Community–Technical College contributed one.

There were four individuals who were members of an advisory board for the project. It was at a meeting of this board that the concept of the consulting firm, later to be named the Innovative Options and New Solutions Corporation, or the I.O.N.S. Corporation, was hatched. The advisory board members included the following individuals (listed in alphabetical order):

John Amend, *Montana State University*

Onofrio Gaglione, *New York City Technical College (retired)*

Paul Kelter, *University of North Carolina/Greensboro*

Kathleen Schulz, *Sandia National Laboratory*

There were six very faithful colleagues who graciously field tested many of the experiments over a period of 1 or 2 years. These folks (listed below in alphabetical order) were very free with both criticism and encouragement and contributed greatly to the manuscript's quality.

Dale Buck, *Cape Fear Community College*

Paul Grutsch, *Athens Area Technical Institute*

Aniruddh Hathi, *Texas State Technical College*

Leslie Hersh, *Delta College*

Joseph Rosen, *New York City Technical College*

Karen Wosczyna-Birch, *Tunxis Community–Technical College*

There were several colleagues who agreed to perform a last-minute detailed review of the experiments and their input also was very important. These include Kelter, Gaglione, Buck, Grutsch, Hersh, and Wosczyna-Birch, who are listed above, and also the following:

Ken Chapman, *American Chemical Society (retired)*

Susan Marine, *Miami University/Middletown*

The grant from the E.I. DuPont DeNemours Company funded a conference, later to be dubbed "The DuPont Conference," at which colleagues from the chemical industry presented their thoughts and ideas as to what elements should be included in an industry-based laboratory manual. The participants' enthusiasm for the project and their advice were extremely important. Following is a list of the participants in alphabetical order.

Deb Butterfield, *Eastman Kodak*

Ed Cox, *Procter and Gamble*

Sue Dudek, *Monsanto*

Ruth Fint, *DuPont*

Charlie Focht, *Nebraska Agriculture Laboratory*

Dennis Marshall, *Eastman Chemical Company*

Dan Martin, *LABSAF Consulting*

Ellen Mesaros, *DuPont*

Jerry Miller, *Eastman Kodak*

Connie Murphy, *The Dow Chemical Company*

Karen Potter, *University of Nebraska*

Richard Sunberg, *Procter and Gamble*

Fran Waller, *Air Products and Chemicals*

Gwynn Warner, *Union Carbide*

The project is indebted to Dan Martin (listed above) who graciously provided the basis for the Safety Manual used in this book. Another important safety resource that proved helpful is the Laboratory Safety Institute, Natick, MA (Web site: http://www.lab-safety.org/).

Others who assisted by reviewing various drafts of experiments along the way include the following:

Clarita Bhat, *Shoreline Community College*

Robert Hofstader, *Exxon Corporation (retired)*

Lynn Melton, *University of Texas at Dallas*

Carol White, *Athens Area Technical Institute*

The author would like to acknowledge Mickey Sarquis and Amy Stander of the Partnership for the Advancement of Chemical Technology (PACT) for allowing me to link two of the experiments to the PACT publication, *Building Student Safety Habits for the Workplace*. I would also like to acknowledge the American Chemical Society's Committee on Chemical Safety. Its *Guidelines for Authors of Laboratory Manuals* provided a framework for communicating the safety issues of each experiment. Many thanks go to the Molecular Arts Corporation for granting permission to use some of their ChemClipArt 1000 images in this book.

Finally, the author is indebted to the program officers in the Advanced Technological Education (ATE) program at the National Science Foundation, especially Dr. Frank Settle, who was a program officer when the project began in 1997, and Dr. Elizabeth Teles, who has kept all of us on our toes since ATE was created in 1994.

Contents

Dedication

*To all the students who have passed through my
laboratory at Southeast Community College since 1977.
Without them, I would not have the job I love
nor the incentive to write books.
May God bless all of you with many precious gifts.*

*"Happy are those who dwell in your house.
They never cease to praise you."* **(Ps 84 5)**

To the Student...

The I.O.N.S Corporation is a fictitious consulting firm created to serve as your employer for some simulated real-world laboratory work to be performed in the college laboratory. You are a laboratory technician for this company and will be working as part of a team, performing laboratory analyses, keeping a notebook, solving problems, making decisions, and writing reports much like a laboratory technician employed by the chemical industry. You will be expected to always exercise good laboratory practices (GLPs) and good management practices (GMPs) as you perform the standard operating procedures (SOPs) necessary to obtain the required results.

The Chief Executive Officer (CEO) of the I.O.N.S. Corporation is Claire Hemistry. Besides serving as CEO, Claire secures all the contracts for the company and always brings a new problem to you via a written memo. She arranges for individuals to serve as industrial and university consultants. You will be in contact with these experts to obtain relevant facts about the work which will help in solving the problem. These consultants will be giving you some initial direction for each problem you undertake. Following is an introductory memo from Ms. Hemistry.

I.O.N.S.
Innovative Options and New Solutions

Memo to: New Employees of I.O.N.S.

From: Claire Hemistry, CEO

Welcome to the I.O.N.S. Corporation! We are very pleased that you have joined us in our efforts to provide quality laboratory and consultation work for our clients. While specific aspects of your job will be communicated to you by your immediate supervisor, there are some very important general expectations to which I would like to call your attention. These relate to our safety policies and procedures and our expectations regarding laboratory notebooks and reporting memos.

First, regarding safety, the I.O.N.S. safety manual is included herein. We are very proud of the fact that only minor safety violations have occurred since our company was founded in 1978. We expect that all safety policies and procedures outlined in the manual will be in effect at all times. For this reason, it is imperative that you diligently study and familiarize yourself with the material contained in this manual. Before you are permitted to do any laboratory work, you must satisfactorily demonstrate your knowledge of safety to your supervisor. He or she will go over the safety manual with you and point out important aspects of your specific workspace. In addition, Ben Whell, the I.O.N.S. Safety Coordinator, provides a safety report for all projects that require special attention with respect to safety.

Second, the lifeline of our company rests solely on our ability to acquire and maintain the confidence of our clients. Accordingly, we expect that all personnel will abide by all GLP procedures set forth by the federal agencies that regulate a given project that you will undertake, such as the Food and Drug Administration (FDA) and the Environmental Protection Agency (EPA). We are a fully certified GLP laboratory and are audited periodically by these agencies. A critical aspect of GLP is good recordkeeping. We cannot emphasize enough the importance of keeping good records as you perform your laboratory work. Accordingly, we have adopted some very basic rules regarding the keeping of notebooks and journals in the laboratory. These are also included herein. We expect you to abide by these rules at all times in all of your work.

Finally, I.O.N.S. workers are asked to write report memos to clients for all projects completed. For each client project you undertake, I will provide the names of the person to whom you should address the memo. Specific instructions and an example reporting memo also are included herein. Both your supervisor and I will evaluate your memos before they are sent.

Again, welcome to the company. If you have any questions at any time, please see your supervisor, or call and set up an appointment with me.

Sincerely,

Claire

Claire Hemistry
CEO, I.O.N.S. Corporation

Section **1**

I.O.N.S. Safety Manual

Innovative Options and New Solutions

Safety Manual Contents

I. Safety Practices

A. Chemical Hygiene Plan

1. All workers must familiarize themselves with and know the location of the I.O.N.S. Chemical Hygiene Plan for the particular laboratories in which they work. The effect of this plan is to define the safety training needed for the laboratory, the circumstances requiring special approval, standard operating procedures for working with hazardous chemicals, criteria for control measures, measures to ensure proper operation of safety equipment, provisions for special hazards, provisions for medical consultants, and the designation of the I.O.N.S. safety hygiene officer.

B. General Practices

1. Horseplay in the laboratory is strictly prohibited.

2. Never climb onto or stand on chairs and stools.

3. Never place bottles, beakers, flasks, etc., containing solutions or chemicals precariously close to the edge of a bench, etc., where they can be knocked off easily.

4. Read and take seriously all signs placed in the area by supervisors or plant personnel, such as "Emergency Exit Only," "No Smoking," "Wet Floor," etc. (Figure 1.1). If you see any potential safety hazard, such as a wet floor, contact your supervisor immediately so that signs may be erected and/or the problem rectified.

FIGURE 1.1
All signs should be taken seriously.

5. If toxic, dangerous, or unpleasant gases become present in the lab at unsafe levels, all personnel should vacate the lab immediately and the plant supervisor notified. Examples include natural gas leaks, an experiment gone awry, etc.

6. Avoid raising chemicals or solutions above eye level.

C. Laboratory Safe Practices

1. All laboratory workers will adopt a safety-first policy. This means you must have adequate **knowledge** and a safe **attitude**.

2. Be cautious and careful, not sloppy and reckless.

3. Smoking, eating, or drinking in the laboratory is not allowed at any time (Figure 1.2).

4. Wear the appropriate personal protection equipment and adopt common sense policies regarding personal safety.

   a. Safety glasses with side shields must be worn at all times (Figure 1.3). (See also Section IIB.)

   b. Latex (or other appropriate material) gloves must be worn if your hands will possibly contact potentially toxic or hazardous materials (Figure 1.4).

   c. Tongs or hand insulators must be used when handling hot materials and equipment.

   d. Shorts, skirts, and open-toed shoes are discouraged.

   e. Tie back long hair, especially when working with flames.

FIGURE 1.2
Eating or drinking in the lab is not allowed at any time.

FIGURE 1.3
Safety glasses with side shields must be worn at all times in the laboratory.

FIGURE 1.4
Wear gloves to avoid contact with hazardous materials.

5. Know the exact location and proper operation of safety equipment, including eyewash station, safety shower, fire extinguishers, fire blankets, fume hoods, first aid kit, etc. Check frequently to make sure that there is nothing obstructing access to safety equipment.

6. Be familiar with the emergency exit route (Figure 1.5). In the event of a fire alarm, proceed in an orderly fashion to the nearest exit.

7. In the event of the sounding of any emergency signal, such as a tornado warning, proceed as directed by the laboratory supervisor.

8. NEVER work alone in the laboratory.

9. Do ONLY the assigned laboratory work as directed by your supervisor.

Emergency
Exit

FIGURE 1.5
Know the emergency exit route.

10. Handle chemicals with CARE.
     a. Read labels to make sure you have the desired chemical.
     b. Keep lids on bottles except when removing chemicals.
     c. Never touch, taste, or inhale a chemical (Figure 1.6).
     d. Wipe up spills immediately following the prescribed clean-up procedure.
     e. Wash hands thoroughly before leaving the lab and anytime you know you have contacted a chemical. Be familiar with the location of a sink with a soap dispenser.

FIGURE 1.6
Never touch, taste, or inhale a chemical.

11. Report all accidents, injuries, spills, or other incidents to your supervisor IMMEDIATELY, no matter how minor.
12. Good housekeeping is conducive to good safety practices. Keep your work area clean and uncluttered. No books, purses, coats, etc., on lab benches.
13. Clean and put away all apparatus, glassware, and equipment before leaving the lab.
14. Dry wet floors immediately to avoid slippery conditions.
15. Thoroughly familiarize yourself with a procedure before beginning and identify potential hazards and safety precautions ahead of time (Figure 1.7).
16. Check glassware and equipment before using. Do not use chipped or cracked glassware or faulty equipment.
17. Avoid touching hot objects. Use tongs or other protective equipment.
18. Be aware of what other laboratory personnel are doing.
19. All samples to be stored or analyzed must be identified as to contents or composition, hazards, name, and date whenever possible.

FIGURE 1.7
Identify potential hazards ahead of time.

FIGURE 1.8
Never pipet by mouth.

20. Never pipet by mouth (Figure 1.8). Use suction devices.

21. Be watchful of broken glass in sinks, drawers, and other places.

22. Consult your supervisor before cleaning glassware with any cleaning agent other than soap and water.

23. Dispose of broken glass in the trash receptacle so designated.

24. If an experimental procedure calls for taking a sample at a remote, off-campus site, only do so if it is safe. Do not, for example, attempt to obtain an environmental water sample if stable footing cannot be maintained.

II. Safety Equipment

A. Emergency Equipment

Emergency equipment is used in case of an unusual happening. Free access to the equipment must be available at all times.

1. *Fire alarms:* A fire alarm must be turned on immediately if there is the remotest possibility that a fire cannot be extinguished with a portable extinguisher.

2. *Fire extinguishers:* The fire extinguishers placed in the laboratories are the CO_2-type. To use the extinguisher, pull the pin, direct the discharge nozzle at the base of the fire, and squeeze the handle (Figure 1.9).

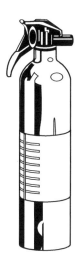

FIGURE 1.9
Pull the pin, point the nozzle, and squeeze the handle.

3. *Fire blankets:* For clothing fires, use the wall-mounted blanket accessible with a loop for inserting your arm. Insert your arm and rotate your body such that the container opens and the blanket wraps around you. For benchtop fires, remove the blanket from the wall-mounted canister through the bottom door, unroll, and cover the fire.

4. *Safety showers:* Safety showers (Figure 1.10) are used when corrosive chemicals are spilled on the body or clothing and in case of fire on the clothing or body. Remove the affected article of clothing before using the shower. The shower also may be used to flush the eyes with water in the event any chemical comes into contact with the eye. Safety showers should be checked regularly for proper operation.

FIGURE 1.10
Safety shower.

5. *Eyewash fountains:* Eyewash fountains (Figure 1.11) are provided for washing the eyes with copious amounts of water. Any clean water source can be used for washing the eyes. A 10- to 15-minute washing time is recommended. Eyewash fountains should be checked regularly for proper operation.

FIGURE 1.11
Eyewash fountain.

6. *Traps in laboratory sinks and drains:* Traps should be checked and filled once a month with water to maintain liquid seal. If corrosive or toxic materials are known to be present, make sure the trap is flushed thoroughly with water. These precautions are taken to prevent backflow of gaseous wastes that may be hazardous.

7. *Fume hoods:* Proper flow of air through fume hoods should be checked periodically.

B. Protective Equipment

1. *Personal equipment* (Figure 1.12)

   a. *Eye protection:* The minimum eye protection designated for the laboratory (American National Standards Institute Standard Z87.1-1989) must be used at all times while in the laboratory **regardless of the activity**.

   b. *Clothing:* Lab coats, coveralls, or uniforms may be required in a given laboratory. These may be required to be flame retardant. If the fire hazard is on the benchtop, a flame retardant lab coat is recommended. If the fire hazard is below the lab benchtop, a flame retardant coverall is recommended.

FIGURE 1.12
Be prepared with eye protection, clothing, aprons, and gloves.

    c. *Aprons:* Rubber or vinyl aprons may be required for some lab work.

    d. *Gloves:* Latex gloves are readily available in the laboratory to protect skin from corrosive and toxic chemicals. Other gloves may be required in some instances, such as for heat protection, or in any case in which latex gloves do not provide adequate protection.

2. *General equipment:* Other equipment, such as face shields (American National Standards Institute Standard Z87.1), bottle and flask carriers, etc., may be required at times. Hood windows should be lowered for control of toxic fumes and when handling otherwise hazardous material in the hood. Assure that a hood is functioning properly by use of an air-flow measurement device.

III. Housekeeping

A. General Philosophy

Housekeeping has an important role in preventing laboratory accidents. Laboratories must be kept in a neat, orderly condition. Maintenance of a clean and orderly workspace is indicative of job interest, personal pride, and safety mindedness; and the training in orderliness is translated into a methodical approach to other lines of endeavor. Good housekeeping in the laboratory is the responsibility of all people working in that laboratory.

B. NFPA Diamond Label

The diamond label in Figure 1.13, called the National Fire Protection Agency (NFPA) diamond provides information at a glance and should be mounted on all containers. Each of the smaller diamonds within the larger one has a different color. The top diamond is red and gives flammability information. The left diamond is blue and tells to what extent the material is a health hazard. The right diamond is yellow and provides reactivity information. The bottom diamond is white and gives special information about the material. The numbers 1, 2, 3, and 4 are placed in the top three smaller diamonds to indicate the severity of the hazard, 1 the least hazardous and 4 the most hazardous.

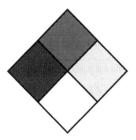

FIGURE 1.13
The NFPA diamond provides safety information at a glance.

C. Arrangement

1. When choosing a site for an operation or the storage of materials, consideration must be given to toxicity, flammability, or other hazardous properties of the material being handled. The information provided by the NFPA diamond is helpful for this.

2. A definite storage place must be provided for each item and the item stored in this place when not in use.

3. All equipment must be clean, dry, and in good repair before being returned to storage.

4. Aisles and walkways must be kept clean, dry, and free of obstructions or tripping hazards.

5. Store wearing apparel in appropriate places. Laboratory aprons must not be hung on the same rack with personal clothing.

D. Equipment

1. *Assembly*

    a. Apparatus must be assembled in a stable, orderly fashion.

    b. Apparatus must not be placed near the edge of the table where there is a possibility of its being accidently knocked by a passerby.

2. *Maintenance*

    a. All defective equipment must be brought to the attention of the laboratory supervisor.

    b. All electrical equipment must be grounded. No unit with frayed cords will be used (Figure 1.14).

    c. Apparatus and working surfaces should be cleaned frequently to prevent accumulations of dust and other foreign matter.

    d. Hood windows should be kept clean.

    e. Sinks must be kept free of debris.

    f. Drawers and cabinet doors must be kept closed when not in use.

DO NOT OPERATE

FIGURE 1.14
Be aware of faulty electrical equipment, such as frayed cords.

E. Spills and Leaks

1. Spilled materials must be cleaned up promptly. Unidentified spilled material must be considered corrosive (Figure 1.15). Notify others in the immediate area and your supervisor of all spills immediately. In case of a spill, take care of yourself first.

FIGURE 1.15
Unidentified spills must be considered corrosive.

2. Should the spilled material be flammable, all flames in the vicinity must be extinguished immediately.

3. A dust pan and brush should be used for cleaning up broken glass. Leather gloves must be worn when picking up any broken glassware.

4. Approved rubber gloves must be used in cleaning up unidentified materials or when there is a possibility of chemical burns.

5. Unusual odors in the atmosphere may indicate an unsafe condition. Report such conditions to your supervisor immediately.

F. Eating, Drinking, and Smoking

1. Eating or drinking is not allowed unless a given room is designated for eating and drinking. Eating or drinking is not allowed in any laboratory. Smoking is allowed only in designated areas in any building.

2. Laboratory equipment, such as beakers or flasks, must not be used for food or drink.

IV. Laboratory Manipulations

A. General

1. Laboratory samples, chemicals, and solutions will not be brought into the office areas.

2. Start the practice of rinsing the unprotected hands several times during the course of the work period to remove unsuspected amounts of irritating chemicals from hands (Figure 1.16). This is to avoid transfer of material from the hands to the face, to books and notebooks, to clothing, etc.

FIGURE 1.16
Start the practice of rinsing unprotected hands frequently.

B. Handling Methods

1. Materials must be stored in such a manner as to minimize all hazards involved. Incompatible chemicals must not be stored side-by-side in one common catch container or shelf.

2. Heavy loads must not be moved without adequate assistance and proper positioning to avoid strain. Extremely heavy equipment should be moved with a cart.

3. Always add a reagent slowly, never "dump" it in. Observe what takes place when the first small amount is added and wait a few moments before adding more.

4. Before pouring a liquid into a vessel with a stopcock (valve) at the bottom, make sure the stopcock is closed.

5. To avoid violent reaction or splattering while diluting solutions, always pour concentrated solutions into water (Figure 1.17), or into less concentrated solutions, while stirring.

6. Never look down the opening of a vessel unless it is empty.

7. Check what your neighbors are doing before lighting a burner. Flames or an operating hot plate or sparking motors should be kept from the vicinity of flammable solvents.

8. Test tubes should be heated gently along the side, not at the bottom, to minimize superheating.

9. Be careful to point the opening of test tubes away from yourself or others.

10. Do not add boiling chips to hot liquids.

FIGURE 1.17
Always add concentrated acid to water when diluting the acid.

11. Do not combine a low boiling liquid with another liquid that is hotter than the liquid's boiling point.

12. Clean up as you work, keeping your table free of chemicals, dirty glassware, etc.

13. When flammable liquids are being transferred to or from a metal container, the container must be provided with an approved electrical ground.

14. Fume hoods must be used whenever corrosive, foul-smelling, or toxic gases are generated by a process.

C. Glassware Handling

1. Broken, chipped, or cracked glassware should never be used or returned to storage (Figure 1.18). If broken beyond repair, discard in trash receptacle designated for broken glassware.

FIGURE 1.18
Broken, chipped, or cracked glassware should never be used.

2. Fire polish or use abrasive cloth to smooth the edges of all glass tubing before use. A wire gauze or file may be used for smoothing edges in cases where heat may damage equipment or glassware.

3. Inserting glass tubing, rods, or thermometers into stoppers is very hazardous and the following precautions must be taken:

   a. Glass tubing and glass rods must not be used unless properly fire polished on both ends.

   b. Make sure the size of the hole matches the size of the tubing, rod, or thermometer.

   c. The glass and stopper must be lubricated with glycerine, water, or other lubricant, unless contamination cannot be tolerated.

   d. Hands must be protected with soft leather gloves.

   e. The stopper must be held between the thumb and forefinger with the palm of the hand parallel to the direction of force.

   f. Consideration must be given to the glass structure in determining the amount of force to apply. Force must be applied slowly.

   g. The glass must be grasped as closely as possible to the point of entry into the stopper.

   h. Force should be applied with a twisting motion.

   i. Never apply force at a bend in the glassware.

   j. If the stopper is in the neck of a flask, remove from flask before attempting to insert or adjust tubing.

    k. Glass or metal tubing should be inserted completely through the stopper with at least 1/4 inch protruding.

4. Full beakers should be supported by grasping around the sides and under the bottom — never over the top (Figure 1.19).

5. Large flasks or beakers must be supported by the base as well as the side or neck when they are full.

6. When heating materials in glassware with an open flame, the glass must be protected from the flame through use of a wire gauze.

7. When placing liquids in bottles that have a positive closure, about 5% of the volume must be reserved as vapor space to allow for expansion with changes of temperature.

8. Hot flasks containing uncondensed vapors or steam must not be stoppered because of the vacuum formed on cooling.

9. No attempt should be made to catch falling glassware.

FIGURE 1.19
Support full beakers or flasks by the base.

V. Material Safety Data Sheets (MSDS)

1. A binder containing all Material Safety Data Sheets of materials present in the laboratory must be on display at all times. This binder must never leave the laboratory.

2. All Material Safety Data Sheets of materials present in the laboratory must be easily accessed when necessary (Figure 1.20).

FIGURE 1.20
MSDSs must be easily accessed in the laboratory.

3. All laboratory workers must be able to clearly interpret Material Safety Data Sheets in their entirety.

4. All laboratory workers must clearly understand handling procedures, hazards, toxicology, and compatibility issues as presented in all MSDSs.

5. All laboratory workers must follow handling procedures as outlined in MSDSs.

VI. Waste Disposal

All laboratory waste (Figure 1.21) must be disposed of according to official procedures of the I.O.N.S. Corporation. These procedures are in compliance with the Resource Conservation and Recovery Act (RCRA) of 1976, the Comprehensive Environmental Response, Compensation, and Liability Act (CERCLA) of 1980, and the guidelines of the Environmental Protection Agency (EPA) and the Occupation Safety and Health Act (OSHA). These procedures are listed in a separate document. All laboratory workers should consult their supervisor for proper disposal procedures following a given laboratory activity.

FIGURE 1.21
Hazardous waste must be disposed of according to law.

Section **2**

I.O.N.S. Laboratory Notebook and Report Memo Guidelines

"All [laboratory] data ... shall be recorded directly, promptly, and legibly in ink."

EPA and FDA GLP Regulations

Contents

Part 1.
Laboratory Notebooks
and Data Recording

This part presents the I.O.N.S. guidelines for laboratory notebooks and data recording. These guidelines contain statements identical to those found in the federal regulations for laboratory notebooks and data recording as given in the Code of Federal Regulations Chapter 21, Part 58, and Chapter 40, Part 160. Chapter 21, Part 58 regulates laboratories governed by the Food and Drug Administration (FDA) and Chapter 40, Part 160 regulates laboratories governed by the Environmental Protection Agency (EPA). These regulations have come to be known as the Good Laboratory Practices regulations or GLP. The GLP statement relating to laboratory notebooks and data recording is given in Figure 2.1.

> All data generated during the conduct of a study, except those that are generated by automated data collection systems, shall be recorded directly, promptly, and legibly in ink. All data entries shall be dated on the date of entry and signed or initialed by the person entering the data. Any change in entries shall be made so as not to obscure the original entry, shall indicate the reason for such change, and shall be dated and signed or identified at the time of the change. In automated data collection systems, the individual responsible for direct data input shall be identified at the time of data input. Any change in automated data entries shall be made so as not to obscure the original entry, shall indicate the reason for change, shall be dated, and the responsible individual shall be identified.
>
> (EPA GLP §160.130(e) and FDA GLP §58.130)

FIGURE 2.1
The GLP statement regarding laboratory notebooks and data recording.

I. General Guidelines

- All notebooks must begin with a Table of Contents. All pages must be numbered and these numbers must be referenced in the Table of Contents. The Table of Contents must be updated as projects are completed and new projects begun (Figure 2.2).

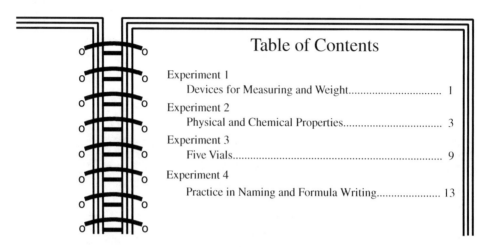

FIGURE 2.2
An example of a Table of Contents.

- All notebook entries will be made in ink. Use of graphite pencils or other erasable writing instrument is strictly prohibited.
- No data entries will be erased or made illegible. If an error was made, a single line is drawn through the entry. Do not use correction fluid. Initial and date corrections and indicate why the correction was necessary (Figure 2.3).

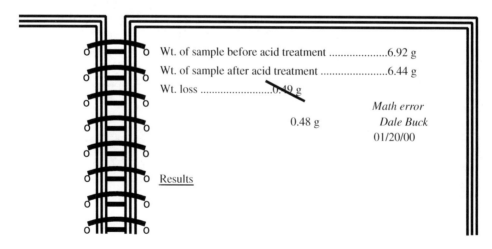

FIGURE 2.3
Initial and date corrections and indicate why the correction was necessary.

- Under no circumstances will any notebook be taken home or otherwise leave the laboratory unless there is data to be recorded at a remote site, such as at a remote sampling site, or unless special permission is granted by the supervisor.

- The following notebook format should be maintained for each project undertaken: (1) Title and Date, (2) Purpose or Objectives Statement, (3) Data Entries, (4) Results, and (5) Conclusions. Each of these are explained below. In each case, write out and underline the words "Title and Date," "Objective," "Data," etc., to clearly identify the beginning of each section (as "Results" is underlined in Figure 2.3).

- Make notebook entries for a given project on consecutive pages where practical. Begin a new project on the front side of a new page. You may skip pages only in order to comply with this guideline.

- Draw a single, diagonal line through blank spaces that consist of four or more lines (including any pages skipped according to the above guideline). These spaces should be initialed and dated.

- Never use a highlighter in a notebook.

- Each notebook page must be signed and dated (Figure 2.4).

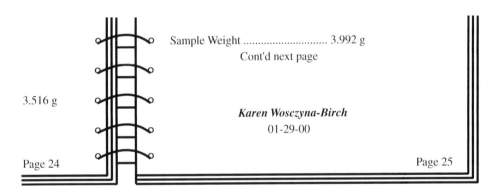

Sample Weight .... 3.992 g
Cont'd next page

Karen Wosczyna-Birch
01-29-00

3.516 g

Page 24

Page 25

FIGURE 2.4
Each notebook page must be signed and dated.

II. Title and Date

- All new experiments will begin with the title of the work and the date it is performed. If the work was continued on another date, that date must be indicated at the point the work was restarted.

- The title will reflect the nature of the work or shall be the title given to the project by the study director.

III. Purpose or Objectives Statement

- Following the Title and Date, a statement of the purpose or objective of the work will be written. This statement should be brief and to the point.

- If appropriate, the SOP will be referenced in this statement.

IV. Data Entries

- Enter data into the notebook as the work is being performed. Entries should be made in ink only.

- If there is any deviation from the SOP, permission must be obtained from the study director and this must be thoroughly documented by indicating exactly what the deviation was and why it occurred.

- The samples analyzed must be described in detail. Such descriptions may include the source of the sample, what steps were taken to ensure that it represents the whole (reference SOP, if appropriate), and what special coding may be assigned and what the codes mean. If the codes were recorded in a separate notebook (such as a "field" notebook), this notebook must be cross-referenced.

- Show the mathematical formula utilized for all calculations and also a sample calculation.

- Construct data "tables" whenever useful and appropriate (Figure 2.5).

Data

Unknown #	Test	Observation	Conclusion
26	Flame test	No distinguishing colors	Na^+, K^+, Li^+, and Ca^{2+} not present
43	Flame test	Bright yellow	Na^+ present
90	Flame test	No distinguishing colors	Na^+, K^+, Li^+, and Ca^{2+} not present
8	Flame test	No distinguishing colors	Na^+, K^+, Li^+, and Ca^{2+} not present
15	Flame test	Bright orange	Ca^{2+} present

FIGURE 2.5
A portion of the data table. In this case, it is one that a technician may have created for Experiment 3. Note the word "Data" underlined to begin the data section.

- Both numerical data and important observations should be recorded.

- Limit attachments (chart recordings, computer printouts, etc.) to one per page. Clear tape or glue may be used. Do not use staples. Only one fold in attachments is allowed. Do not cover any notebook entries with attachments.

V. Results

- The results of the project, such as numerical values representing analysis results, should be reported in the notebook in table form if appropriate. Otherwise, a statement of the outcome is written, or if a single numerical value is the outcome, then it is reported here. In order to identify what is to be reported as results, consider what it is the client wants to know.

VI. Conclusion

- After results are reported, the experiment is drawn to a close with a brief concluding statement indicating whether the objective was achieved.

Part 2.
Guidelines for the
Report Memo

This part provides the I.O.N.S. guidelines for the memos that individuals who perform the work will write to the clients. The word "memo" is short for memorandum. In general, a memo is meant to be a brief and to-the-point written document that reports information and/or recommends action. Memos are designed for efficient distribution and easy, quick reading. For I.O.N.S., this means that we will report our results to our clients in as efficient a manner as possible. It will be a brief, to-the-point presentation for easy, quick reading. Our major objective with the report memo is the satisfaction of our customers. Such is the task of analytical scientists throughout the industry, as conveyed in the quotation in Figure 2.6, and is certainly the task of analytical scientists in the I.O.N.S. Corporation.

> "Industrial scientists and engineers spend a significant portion of time writing periodic reports that summarize their work, proposals justifying the purchase of instruments, or suggesting research programs, development plans, and appraisals for themselves and their subordinates, analytical reports, and memos. Each document must be written to convey conclusions, recommendations, and technical information to the target audience in ways that they clearly understand."

FIGURE 2.6
It is important for analytical scientists to communicate effectively. (From Thorpe, T. and Ullman, A., Preparing analytical chemists for industry, *Anal. Chem.*, 68, 477A–480A, 1996. With permission.)

I. General Guidelines

- Official I.O.N.S. stationery will be used for all memos. This stationery is available from all supervisors.
- If more than one page is needed, plain, white paper will be used for the second and all subsequent pages.
- All memos will be typed.
- All one-page memos will be centered on the page.
- There should be no misspellings, grammar errors, punctuation errors, incorrect word usage, etc.

II. The Heading

- The heading format shall be as follows:

  Date:_____

  Memo to: _____

  From: _____

  Subject: _____

- The "Memo to" line shall give the full name of the client followed by his/her company name.
- The "From" line shall give the full name of the writer, followed by a comma, followed by the words "The I.O.N.S. Corporation."
- The "Subject" line should provide a *brief* subject title that tells the client what the memo is about. A few words that clearly and accurately reflect the overall message should be carefully chosen.

III. The Content

- Begin with a paragraph that identifies the purpose or subject of the memo and provides sufficient background so that the client will have a clear idea of what information is coming in the rest of the memo. Here, the client should be reminded as to the nature of the problem given to I.O.N.S. to solve. A good opening sentence begins as follows: "The purpose of this memo is to report the results of the work of the I.O.N.S. corporation relating to…"
- Next, tell what was tested, why it was tested, how it was tested (without going into great detail), and what the tests showed. You should not provide any raw data, but perhaps indicate to the client that such data are available if he/she would like to see it.
- Indicate what conclusions you came to as a result of the work, why you came to these conclusions, and what are your recommendations. Be careful with recommendations, however, as many things must be left up to the client, such as whether to sue someone, whether to fire someone, whether to spend money to improve a process, etc. We don't want to recommend something that is none of our business.

IV. Closing Signature

- It is the official policy of the I.O.N.S. Corporation for the memo writer to sign the memo at the bottom following the last paragraph. First, the word "Sincerely" is typed, followed by the signature. Following the signature, the writer's full name shall be typed.

V. Example

- An example memo is provided in Figure 2.7.

I.O.N.S.
Innovative Options and New Solutions

January 25, 2000

Memo to: Dr. Paul Grutsch, the Grutsch Chemical Company

From: Joseph Rosen, the I.O.N.S. Corporation

Subject: Beef Cattle Illness

Dear Dr. Grutsch:

The purpose of this memo is to report the results of the work of the I.O.N.S. Corporation relating to the illness that has stricken the beef cattle in your state. Thank you for coming to us with this interesting problem. We do have some important results to report.

We tested the urine samples that you provided, since our consultants agreed that if there is any toxic chemical in the system of the beef cattle, it would likely be found in the urine. The samples were tested by a gas chromatography technique we found that was specifically for identifying toxic chemicals in bovine urine. The results clearly showed that the urine had a very high level of colchicine, a poisonous alkaloid sometimes used to enhance plant growth. We would be happy to meet with you to show you the data that led us to this conclusion.

Since colchicine is toxic to cattle at the level we found, we are certain that this is what caused the illness. We respectively recommend a quarantine of the infected cattle until the levels dissipate. We would be happy to continue to monitor the urine of the quarantined specimens if that would help.

Sincerely,

Joseph Rosen

Joseph Rosen, I.O.N.S. Corporation

FIGURE 2.7
An example client memo.

I.O.N.S. New Employee Orientation

Innovative Options and New Solutions

Contents

Experiment 1
Devices for Measuring Volume and Weight

Introduction

Work in a chemistry laboratory involves a great deal of observation. When we want to know what some of the physical and chemical properties of substances are, we make observations. When we notice that a material substance is in the liquid phase, or that it is red, or that it smells like ammonia, or that it is a free-flowing granular solid, we are making observations concerning its physical properties. When we notice that a clear colorless solution held in a test tube changes to a pink color or causes a white precipitate to form or causes the bubbles of a gaseous substance to form when a second clear, colorless substance is added, we are making observations of chemical change that allow us to make conclusions concerning its chemical properties.

Very often, if we want to observe something quantitative about a material substance, we make measurements along with the observations. For example, if we want to determine what effect of a certain quantity of one material substance has on a certain quantity of another, we must measure the two quantities. If we want to know how heavy a material is relative to its volume, we make measurements of weight and volume on a quantity of the material. If we want to know the temperature at which a certain liquid boils, we measure the temperature of the liquid when it is boiling. Even the most advanced chemistry professional working in his/her laboratory is constantly making observations and measurements on material substances.

Probably the most basic of all measurements in a chemistry laboratory are those of mass and volume. **Mass** is a measure of the quantity of matter as indicated by its **weight**, or the effect of the gravitational force on the mass. For example, the mass of our body is measured whenever we step onto a bathroom scale. **Volume** is a measure of how much space a quantity of matter occupies. In the kitchen, when a recipe calls for two cups of milk, the reference to "two cups" is a reference to the volume of the milk needed.

In a chemistry laboratory, the metric system of measurement is used rather than the more familiar English system. The most common measurement unit for mass is the gram, not the pound. The most common measurement unit for volume is the milliliter, not the cup. One milliliter of water is a little more than half a tablespoon of water. One milliliter (abbreviated "mL") of water weighs approximately one gram (abbreviated "g").

Balances

The device for measuring mass in a chemistry laboratory is called a **balance**, not a **scale**. Balances in the modern laboratory are almost always digital. This means that the mass reading is ascertained by simply observing a series of numbers in a digital display. Thus, such devices are very easy to use. It is often just a matter of placing the object to be weighed on the balance "pan" and observing the reading in this display. However, different balances found in a laboratory often have different capabilities in terms of capacity and precision. **Capacity** refers to the maximum mass that can be read. On some balances, a chemist cannot measure more than 100 g. On others, he/she can measure up to 500 g. On still others, extremely large weights can be measured. **Precision** refers to the degree of reproducibility of a measurement or how well several measurements on the same object agree. Precision is reflected in the number of digits that can be observed in the reading. Some balances have a precision of ±0.1 g, meaning that beyond the first decimal place, the measurement is no longer reproducible. Others have a precision of ±0.01 g, and still others have a precision of ±0.0001 g. Some balances are known as **top-loading balances** because the pan is on the top of the balance. These often have a precision of ±0.1 g or ±0.01 g. When more precision is required, the pan may be enclosed in a glass enclosure with sliding doors. This is so that air currents will not disturb the weighing. These balances, known as **analytical balances**, have a precision of ±0.0001 g or even ±0.00001 g. Drawings of these balances are shown in Figure 3.1 and examples of the digital displays are shown in Figure 3.2. All balances must be zeroed (made to read 0 g when there is nothing on the pan) before the measurement is made.

Top-Loading Balance

Analytical Balance

FIGURE 3.1
Balances.

68.5 g   68.51 g   68.5131 g

FIGURE 3.2
Digital displays for several balances. Left, a top-loader readable to the first decimal place. Center, a top-loader readable to the second decimal place. Right, an analytical balance readable to the fourth decimal place.

Graduated Cylinders, Burets, Measuring Pipets, and Pipetters

Quantities of liquids, such as water and water solutions, are often measured in the laboratory by measuring their volumes. The chemist contains them in a piece of glassware with graduation lines, much like the graduation lines on the measuring cup in your kitchen. Typical examples are graduated cylinders, burets, and measuring pipets (Figure 3.3). Beakers and flasks are *not* used for precise volume measurements, even though they may have graduation lines. Such lines are only meant to be crude indications of volume. The devices in Figure 3.3 are, by design, not digital. Rather, the measurement is made by reading the graduation lines where the **meniscus** (curved surface) of the liquid is found. The number of digits (precision) is determined by writing down all the digits you know with certainty followed by one that is estimated. The fact that the meniscus is curved presents a special problem. Where on the curved surface is the reading to be made? The answer is that the glassware is calibrated at the factory such that the position of the *bottom* of the meniscus is what is observed relative to the graduations lines. Examples are given in Figure 3.4. Notice that for burets and measuring pipets, the position of the meniscus is determined *after* the volumes are dispensed from the bottom, while for graduated cylinders the meniscus is read *before* the volume is dispensed by inverting and delivering from the top. The graduation lines on burets and pipets, therefore, are read from the top down, while on graduated cylinders they are measured from the bottom up (see Figure 3.4).

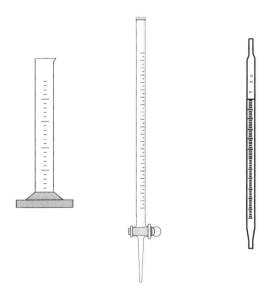

FIGURE 3.3
Glassware for measuring volumes of liquids. Left to right, graduated cylinder, buret, and measuring pipet.

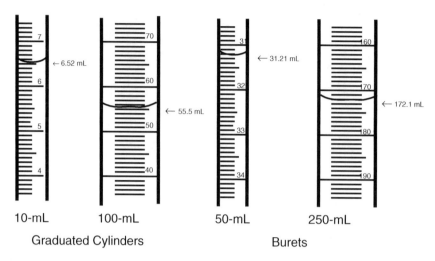

10-mL   100-mL   50-mL   250-mL

Graduated Cylinders   Burets

FIGURE 3.4
Several examples of meniscus readings.

Another volume measuring device that is in common use is called the pipetter. A pipetter is a hollow plastic device with a "push button" at the top and a removable, disposable plastic tip, in which the liquid is contained, at the bottom (Figures 3.5). Enclosed by the plastic barrel is a rubber bulb that is evacuated by pressing down on the button. The volume to be measured is read and "dialed in" at the push button. The reading is digital. The device is used by first dialing in the desired volume on the digital display. This is followed by pushing the push button to a "stop." A bottom part of the tip is immersed and the button released slowly so that the desired volume is drawn into the tip. This volume is then dispensed by pressing down on the push button past the first "stop" to a second "stop." This ensures that the liquid is totally dispensed by blowing out the last droplet held in the tip. The volumes dispensed by pipetters are typically read on the digital display in microliters because these volumes are usually quite small. A microliter (abbreviated µL) is a factor of a thousand smaller than the milliliter. Thus, a reading of 165 µL is the same as 0.165 mL.

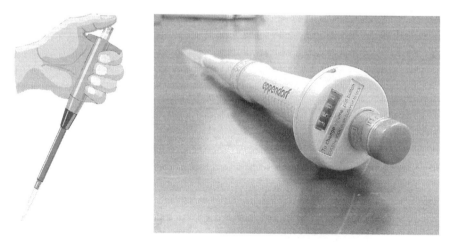

FIGURE 3.5
Two different views of a pipetter: (left) the pipetter with left thumb on the push button; (right) a top view of the pipetter showing the digital display.

Overview of Activity

In this laboratory activity, new I.O.N.S. employees are asked to practice making measurements of mass and volume using the various devices described above. You will be measuring the mass of an object on three different laboratory balances and making sure that the precision of your reading matches the precision of the balance. You also will be making readings on several graduated cylinders and burets that have been set up in the laboratory. In addition, you will be comparing the volumes and masses of quantities of water measured by different devices and methods, including counting drops.

Procedure

Note: *Please follow the I.O.N.S. policies and procedures regarding laboratory notebooks while performing this exercise.*

1. Your supervisor has identified three balances for your use in this exercise and has labeled them "A", "B", and "C". Using gloves to avoid fingerprints, measure the mass of a dry 25-mL beaker on each of these balances. Record these masses in your Data section and label clearly.

2. Deliver 4 mL of water to the beaker using a 10-mL graduated cylinder, making sure that the bottom of the meniscus rests on the 4-mL line of the cylinder before delivery. Now measure the mass of the beaker again on each of the three balances and subtract the mass of the empty beaker so as to obtain the mass of the water for each balance. Record all mass measurements in your Data section, including the mass of the beaker with the water in it and the mass of the water alone. There will be three sets of data, one for each balance. Clearly label all entries.

3. Empty the beaker and dry it as well as possible with a paper towel. Place 4 mL of water in the beaker again, but this time measure this volume by adding the water from a 50-mL buret. Again measure the mass of the beaker on the three balances and subtract the mass of the empty beaker to obtain the mass of the water on each balance. Then pour the water into the graduated cylinder and record the volume reading. Be sure to record the volume measurement and all mass measurements in your Data section and clearly label all entries. Again you will have three sets of data for the mass measurements, one for each balance.

4. Repeat Step 3 a number of times, measuring the 4 mL of water by a different method each time. Possible methods include:

    a. A pipetter

    b. A 10-mL buret

    c. A 250-mL buret

    d. A glass dropper, assuming that 20 drops is 1 mL

    e. A disposable plastic dropper, again assuming that 20 drops is 1 mL

    f. A glass dropper with a very narrow bore, such as one found on some dropper bottles, assuming 20 drops is 1 mL

    g. A plastic dropper bottle designed to dispense drops through a small hole in a plastic stopper, also assuming that 20 drops is 1 mL

    Your supervisor will assign specific methods according to availability of droppers and dropper bottles in your assigned laboratory.

5. In the Results section of your notebook, analyze your data and comment on the usefulness of each kind of balance for this exercise and also on the legitimacy of measuring 4 mL of water with each method used. Point out specific data entries that support your comments.

Experiment 2
Physical and Chemical Properties: Identifying Ordinary Household Products

Introduction

This experiment is an exercise in the observation of the physical and chemical properties of substances. The substances chosen are consumer products manufactured for use in and around the home, or "household products," each a white solid. Since physical and chemical properties are characteristics that can identify a substance, the observations made in this experiment are used for that purpose — identification.

The laboratory procedures associated with identification comprise a process known as "qualitative analysis." In Part A of this experiment, you will examine 11 household products to determine some specific physical and chemical properties and to establish a qualitative analysis scheme for their identification. The physical properties include characteristics associated with their physical appearance, including color, particle size, and texture, as well as their solubility in water, rubbing alcohol, and hot water. The chemical properties include the manner in which the white solids react chemically with various other chemicals. In Part B, you will have three to five unknown household products (taken from those tested in Part A) and your work will involve using your qualitative analysis scheme to identify them. The flow chart that follows the procedure (Figure 3.6) should be filled out while performing Part A and should help when you perform Part B.

The 11 white, solid household products used in this experiment are table salt, baking soda, washing soda, drain opener, boric acid, plaster of Paris, calcium supplement (calcium citrate), cornstarch, fruit sugar, table sugar, and epsom salt.

I.O.N.S. Safety Report
prepared by Ben Whell, I.O.N.S. Safety Coordinator

Experiment 2

Equipment and Technique

- Use small plastic dropper bottles to contain reagents (rubbing alcohol, 0.30 M NaOH, vinegar, phenolphthalein, and Benedict's Reagent) to avoid breakage and large spills.
- Use test tube racks so that test tubes will not slip out easily.
- Do not point the open end of a test tube at another person.
- Use caution with hot water bath (hot beaker on hot plate) to avoid burns. Do not use a hot plate that has a frayed cord.
- Use special care when shaking hot test tubes so that the hot solution stays contained.

Chemicals

- Handle drain opener with care. Do not allow it to contact skin or clothing.
- Keep tincture of iodine in original container purchased at pharmacy.

Workplace Cleanup

- Contents of all test tubes and spot plates can be discarded down the drain.
- Test tubes, spot plates, spatulas, etc., should be rinsed with water and dried with a towel before storing.
- Plastic reagent bottles and all sample vials should be placed in appropriate storage area.
- Latex gloves should be discarded in trash receptacles.
- Switch off hot plates and allow to cool before storing.

Hazards Classifications

- Potential injuries include cuts from broken test tubes, burns from contacting hot water or hot plate, chemical burns from drain opener, etc. These are treatable on site.

Laboratory Safety Quiz

1. What should you do if you weren't wearing gloves and you accidentally contacted drain opener with your fingers?
2. How does the handling of reagents with plastic dropper bottles help with safety hazards?
3. Name some specific scenarios that might occur in the procedure in which safety glasses would provide a measure of protection.
4. If you burned your finger on a hot plate and this caused you to drop and break a test tube filled with a solution, what should you do?

Procedure

Notes: (1) You should work as part of a team of two to four analysts to discover the properties in Part A, but work on your own to analyze the unknowns in Part B. (2) Whenever you are directed to "shake" a test tube, wear latex gloves and place your finger over the mouth of the tube and then shake the tube. Before proceeding to the next test tube, rinse your latex-covered finger with distilled water to avoid cross-contamination. (3) See the "Suggested format for lab notebook" at the end of the procedure. (4) Use distilled water whenever the procedure calls for water. Reminder: Safety glasses are required and latex gloves should be worn.

Part A. Determination of Properties

1. Examine the 11 white solids which have been placed in labeled containers. Record, in your notebook, any physical properties that are unique and may distinguish a given white solid from the others. Examples of such properties are color distinctions (off-white, shiny-white, etc.), particle size (larger crystals, powdered, etc.), and texture (rough, smooth, etc.).

2. Test the solubility of each white solid in water. To do this, perform the following series of steps.

   a. Line up 11 clean, empty test tubes in a test tube rack and label each with the name of a white solid (table salt, baking soda, drain opener, etc.).

   b. Remove a *small* portion (less than half the size of a pea) of each white solid with a spatula and place each in the test tube with the matching label.

   c. Add 5 mL of water to each of the 11 test tubes and then shake each tube according to the directions in Note 2. Indicate in your notebook whether the solid is soluble or insoluble in water. *Do not discard the contents of any of the test tubes until directed to do so below.*

   d. Complete the flow chart (Figure 3.6) for Step 2.

3. Perform the following test on the contents of all test tubes in which the solid in Step 2 did not dissolve. There should be three such test tubes. Add a drop or two of tincture of iodine and wash it into the solution with a little water. Shake as previously directed. A deep blue color should form in one of the tubes. Record in your notebook which solid gives this result. The deep blue color is unique to this solid and will identify it. The contents of the three test tubes used in this step may now be discarded. Complete the flow chart (Figure 3.6) for Step 3.

4. Next, test the two remaining white solids that did not dissolve in water in Step 2 for their reaction with vinegar. Obtain a porcelain spot plate and place a small amount (again less than half the size of a pea) of each of these two solids in separate wells on the plate (make sure you know which is which). Add a few drops of vinegar. Vigorous bubbling/fizzing should occur with one of the solids and less vigorous or no bubbling/fizzing with the other. Record on the data sheet which solids give which result. The combination of insolubility in water and the vigorous bubbling/fizzing upon contact with vinegar is unique to one of the solids. Also record in your notebook which solid did not give vigorous bubbling/fizzing with vinegar. The combination of insolubility in water and less vigorous or no bubbling/fizzing with vinegar is unique to this solid. Complete the flow chart (Figure 3.6) for Step 4.

5. To each of the remaining test tubes from Step 2, add several drops of phenolphthalein solution and shake. Two of the solutions should turn pink. The solids that you dissolved in these two solutions can be differentiated with vinegar as were the two solids in Step 4. Thus, again use the porcelain spot plates and test the original white solids with vinegar as in Step 4. One will produce bubbling/fizzing and one will not. Record in your notebook which produces bubbling/fizzing and which doesn't. The combination of solubility in water, a pink color with

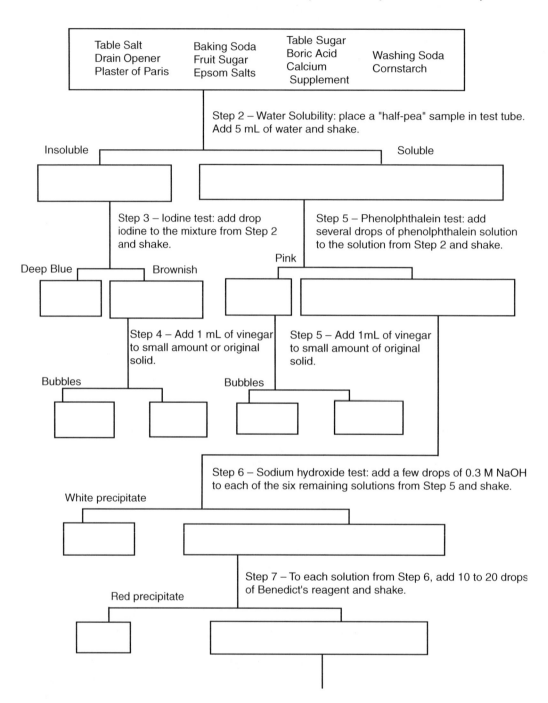

FIGURE 3.6
Flow chart for Experiment 2.

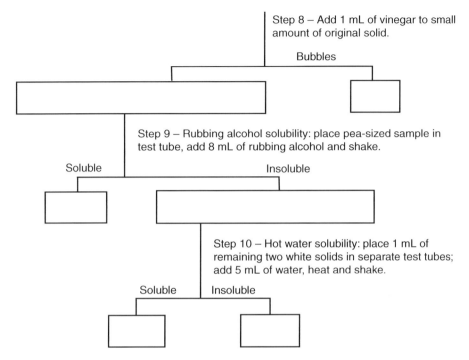

FIGURE 3.6 (cont.)

phenolphthalein, and bubbling/fizzing or no bubbling/fizzing with vinegar identifies these solids. The bubbling/fizzing that occurs is due to the acid in vinegar reacting with carbonates in the solids to produce carbon dioxide gas bubbles. Complete the flow chart (Figure 3.6) for Step 5.

6.   Test the six solutions (the colorless ones) remaining from Step 5 with the solution labeled "0.3 M NaOH." To do this, add a few drops of this solution to each of the six test tubes and shake. All solutions will turn pink and a white precipitate (an insoluble solid) should form in one of the test tubes. This solid may appear as a cloudiness that is difficult to see. If you see no cloudiness in any of the test tubes, add a few more drops of the 0.3 M NaOH to all tubes and shake. Record which white solid produces the white cloudiness and how many drops of the NaOH were required. The white precipitate identifies this solid. Complete the flow chart (Figure 3.6) for Step 6.

7.   Test the five solutions remaining from Step 6 with a solution called "Benedict's Reagent." To do this, add 10–20 drops of Benedict's Reagent to each of the solutions and shake. Heat for 2 or 3 minutes in a hot water bath. A reddish-brown precipitate should form in one of the tubes. This test identifies the original white solid dissolved in this tube. Record what solid this is and complete the flow chart (Figure 3.6) for Step 7.

8.   Test the remaining four unidentified solids with vinegar as in Step 4. Only one of these solids will give bubbles and this then identifies this solid. Record what solid this is in your notebook. Complete the flow chart (Figure 3.6) for Step 8.

9.   Test the remaining three solids for solubility in rubbing alcohol. Do this in the manner described in Step 2 for water, but use 8 mL of the rubbing alcohol instead of 5 mL of water. Only one of the three solids will dissolve and, thus, this test will identify it. Record what solid this is and complete the flow chart (Figure 3.6) for Step 9.

10.   The final two solids can be identified by testing their solubility in hot water. Estimate the 1 mL level for water in two separate test tubes (mark this level) and place one white solid in one test tube and the other in the second tube, each to the 1 mL mark. Then add 5 mL of water. Next, heat the test tubes by immersing them in hot water for a few minutes and then shake. One of the solids will dissolve under these conditions and the other will not. Record which solid dissolves and which doesn't and complete the flow chart (Figure 3.6) for Step 10.

11.   Compare your flow chart with one posted by your supervisor. If there are any discrepancies, repeat the appropriate steps and record your observations.

12.   Discuss which properties you observed in Part A are physical properties and which are chemical properties.

Part B. Identifying an Unknown White Solid

1.   You will be given three to five white solids to identify using the tests learned in Part A. They will be labeled with a letter or letters that no other analyst will have. Record these labels in your notebook. Next, observe the physical appearance of each of your white solids as in Step 1 of Part A and record in the notebook. You may be able to speculate as to what your white solids might be just from these observations. Record these speculations in your notebook. You will then have some expectations when you actually run the tests.

2.   Perform the tests on your unknown solids and record the results in your notebook along with your conclusions as to their identities. The flow chart (Figure 3.6) provides abbreviated procedures and may prove useful.

Suggested Format for Data Section of Lab Notebook Write-Up

Part A

1. *Physical Characteristics.* Create a two-column table with the following headings:

   Solid Name   **Unique Physical Properties**

2. *Water Solubility Test.* Create a two-column table with the following headings:

   Solid Name   **Soluble or Insoluble**

3. Name of solid giving deep blue color with iodine: _____

4. Name of solid giving vigorous bubbling with vinegar: _____
   Name of solid not giving vigorous bubbling with vinegar: _____

5. Name of solids whose solutions give pink color with phenolphthalein: _____
   and _____
   Name of solid giving bubbles with vinegar: _____
   Name of solid not giving bubbles with vinegar: _____

6. Name of solid giving white precipitate: _____

7. Name of solid giving reddish-brown precipitate with Benedict's Reagent: _____

8. Name of solid giving bubbles with vinegar: _____

9. Name of solid dissolving in rubbing alcohol: _____

10. Name of solid dissolving in hot water: _____
    Name of solid not dissolving in hot water: _____

Part B

1. Unknown letters: _____
   Physical Characteristics. Create a three-column table with the following headings:

   Identifying Letter   **Physical Characteristics**   **Speculation**

2. Create a four-column table with the following headings. This table may take up several pages in a notebook, assuming that each test performed on each white solid (up to five) is recorded. Fill in the "Name of Solid" only when the test being recorded resulted in its identification.

   Identifying Letter   **Name of Test**   **Results**   **Name of Solid**

Acknowledgment

This experiment was adapted from the *Journal of Chemical Education,* 68(4), 328–329, 1991. With permission.

Reference

Kenkel, J., Kelter, P., and Hage, D., *Chemistry: An Industry-Based Introduction,* CRC Press/Lewis Publishers, Boca Raton, FL, 2000, Chap. 1, Sec. 1.2.

Experiment 3
Five Vials: Identifying Dissolved Cations and Anions

Introduction

In this experiment you will be given five vials, each filled with a water solution of an ionic compound. Since an ionic compound consists of one positively charged ion, or cation (pronounced cat´-eye-un), and one negatively charged ion, or anion (pronounced ann´-eye-un), each vial will contain one kind of cation and one kind of anion. It will be your job to "analyze" each solution to determine what the cation is and what the anion is in each, a total of 10 ions.

Before beginning your work on the five vials, however, you must learn what a "positive" laboratory test for each cation and anion looks like. To do this, we have made solutions available that contain the ions which are possibilities for the unknown solutions. The cations are Na^+, K^+, Li^+, Ca^{2+}, NH_4^+, Fe^{3+}, Ni^{2+}, Al^{3+}, and Ba^{2+}. The anions are Cl^-, Br^-, I^-, CO_3^{2-}, SO_4^{2-}, NO_3^-, and PO_4^{3-}. You will proceed to perform laboratory tests, such as flame tests, and mixing small amounts of these solutions with small amounts of certain other solutions, and make observations of color, odor, precipitates formed and gases evolved, etc., which characterize each of the ions. This is Part A (cations) and Part B (anions) of the experiment. Part C is the analysis of the solutions in the five vials.

The observations that should be made in each step are noted in the procedure with some exceptions. These observations do not have to be re-recorded in your notebook unless that is your wish. The exceptions, however, should be noted. The pages following the procedure provide a format for these exceptions.

You will be part of a team working on Parts A and B. For Part C, you should design your own experiment and proceed on your own.

I.O.N.S. Safety Report
prepared by Ben Whell, I.O.N.S. Safety Coordinator
Experiment 3

Equipment and Technique

- Use small plastic dropper bottles to contain reagents (acids, bases, solvents, and chemicals to be added to knowns and unknowns) to avoid breakage and large spills.
- Use test tube racks so that test tubes will not slip out easily.
- Do not point the open end of a test tube at another person.
- Use caution with centrifuge. Be sure centrifuge is properly balanced. Switch to "off" if vibration occurs. Do not use a centrifuge with a frayed cord.
- The atomic absorption unit should not be operated without proper supervision. Operate according to instructions and use special caution with flame and in handling acetylene. Be sure the fume hood over the instrument is on and the drain line is properly configured.
- Use caution with hot water bath (hot beaker on hot plate) to avoid burns. Do not use a hot plate that has a frayed cord.
- Use special care when shaking hot test tubes so that the hot solution stays contained.
- Use care when performing the flame test to avoid burns.
- If your supervisor asks you to prepare the chlorine water, do so in a fume hood and keep the fume hood on for the entire period, even after cleanup. Use caution — the odor of chlorine can be quite strong.

Chemicals

- While some chemicals used are potentially hazardous, microscale amounts are used. Consult your supervisor for special considerations regarding safety.
- Vacate the lab if the odor of chlorine from the preparation step becomes too strong.
- Hexane is extremely flammable.

Workplace Cleanup

- While some chemicals used are potentially hazardous, microscale amounts are used. Consult your supervisor for special considerations regarding disposal.
- Test tubes, etc., should be rinsed with water and dried with a towel before storing.
- Plastic reagent bottles and all sample vials should be placed in appropriate storage area.
- Latex gloves should be discarded in trash receptacles.
- Switch off hot plates and allow to cool before storing.
- If spectrophotometers are located in a separate room, remember to clean up.
- Do not dispose of hexane down the drain. Pour from test tubes into hexane waste storage container.
- Solutions containing silver and silver precipitates should not be discarded down the drain. Place in silver waste storage container.

Hazards Classifications

- Potential injuries include cuts from broken test tubes, burns from contacting hot water or hot plate, chemical burns from acids, etc. These are treatable on site if not too severe.
- Inhalation of chlorine and hexane vapors can occur if special care is not exercised.

Laboratory Safety Quiz

1. Why should the chlorine generation apparatus be given special attention?
2. Give several reasons why the use of microscale amounts of chemicals is a good thing.

Procedure

Note: *Use the same procedure for shaking test tubes as you did in Experiment 2. Wear latex gloves, place your finger over the mouth of the test tube and shake. Then rinse your latex-covered finger with distilled water before shaking any other test tube. Reminder: Safety glasses are required.*

Part A. Tests for Cations

Test for Fe³⁺ ions

1. Note the color of the solution labeled Fe^{3+}. This color is due to the presence of the Fe^{3+}. Place one drop of this solution in a test tube, add one drop of 12 M HCl and dilute to approximately 1 mL with water. Shake. Add 1 drop of 1 M KCNS and shake again. A blood-red color is the observation for a positive test for Fe^{3+}. The reaction is

$$Fe^{3+} + KCNS \rightarrow FeCNS^{2+} + K^{+}$$

The $FeCNS^{2+}$ gives the solution the blood-red color.

Test for Ni²⁺ ions

2. Note the color of the solution labeled Ni^{2+}. This color is due to the presence of the Ni^{2+}. To 1 drop of this solution, add 5 mL of water and 10 drops of dimethylglyoxime solution. Shake. Now add two to three drops of concentrated ammonium hydroxide and shake again. The formation of a scarlet red precipitate is a positive test for Ni^{2+}. The reaction is

$$Ni^{2+} + 2C_4H_8N_2O_2 \rightarrow Ni(C_4H_7N_2O_2)_2 + 2H^{+}$$

The $Ni(C_4H_7N_2O_2)_2$ is the scarlet red precipitate.

Tests for Al³⁺, Ba²⁺, and Ca²⁺ ions

3. In three separate test tubes labeled Al^{3+}, Ca^{2+}, and Ba^{2+}, place 1 mL of the solution labeled Al^{3+}, 1 mL of the solution labeled Ca^{2+} and 1 mL of the solution labeled Ba^{2+}. Add 2 drops of 6 M HNO_3 to each, shake, then add 1 mL (20 drops) of 6 M NH_4OH to each and shake again. A gelatinous white precipitate forms in the test tube containing the Al^{3+} ion. This is the positive test for Al^{3+}. No precipitate forms in the other tubes. The reaction is

$$Al^{3+} + 3NH_4OH \rightarrow Al(OH)_3 + 3NH_4^{+}$$

The $Al(OH)_3$ is the gelatinous precipitate.

4. To the two test tubes in which no precipitate formed in Step 3, add 2 mL of 1 M $(NH_4)_2CO_3$. Precipitates should form in both tubes. Centrifuge, carefully withdraw the supernatant (the liquid above the precipitate), and discard. Add 1 mL of water to each, shake, then add 6 M HCl dropwise, shaking after each drop, until the solid has dissolved. Then add 1 mL of 1 M Na_2SO_4 solution and shake. A white precipitate forms in the case of the Ba^{2+} but not in the case of the Ca^{2+}. These observations constitute the positive tests for these ions. The reactions are

$$Ba^{2+} + Na_2SO_4 \xrightarrow{ACID} BaSO_4 + 2Na^{+}$$
$$Ca^{2+} + Na_2SO_4 \xrightarrow{ACID} \text{No reaction}$$

The $BaSO_4$ is the white precipitate.

Flame tests for Na+, K+, Ca2+, and Li+ ions

5.   An instrument known as an atomic absorption (AA) spectrophotometer will be used for this
      test. Ask your supervisor to assist when you are ready for this test. Your supervisor will
      aspirate each into the flame of the AA while you observe the color of the flame. Record in
      your notebook a description of the color given to the flame by each of the ions. These colors
      are characteristic of these ions and will identify them.

Test for the ammonium ion, NH_4^+

6.   When a solution of the ammonium ion contacts a solution of NaOH, ammonia gas, NH_3,
      forms according to the following equation:

$$NH_4^+ + NaOH \rightarrow NH_3 + Na^+ + H_2O$$

Place two drops of the solution labeled NH_4^+ on a watch glass. Add five drops of water and
swirl to mix. Then add 2 drops of 4 M NaOH. Try to detect the odor of ammonia. If you
cannot smell the ammonia, warm the watch glass gently on a hot plate and try again.

Part B. Tests for Anions

Test for phosphate ions, PO_4^{3-}

1.   Set up a hot water bath by half-filling a 250-mL beaker with distilled water and placing it
      on a hot plate so that the water becomes very hot before performing this test. Place five drops
      of the solution labeled "PO_4^{3-}" in a test tube and add two drops of concentrated HNO_3. Shake.
      Add five drops of ammonium molybdate solution, shake, and place in the hot water bath for
      5 minutes. The formation of a yellow color or yellow precipitate is the positive test for
      phosphate. Bear in mind that contamination with residue from phosphate-containing soap
      may give a false-positive test for the unknowns.

Test for carbonate ions, CO_3^{2-}

2.   Place 1 mL of the solution labeled "CO_3^{2-}" in a test tube. While carefully watching as the
      drops contact the solution, add 2 drops of 1.5 M H_2SO_4 and check for the formation of gas
      bubbles. Carbonate will form carbon dioxide gas when acidified and these bubbles will be
      clearly visible for a fraction of a second as the acid contacts the solution. The reaction is

$$CO_3^{2-} + H_2SO_4 \rightarrow CO_2 + H_2O + SO_4^{2-}$$

Test for sulfate ions, SO_4^{2-}

3.   Place 1 mL of the solution labeled "SO_4^{2-}" in a test tube and add 1 mL of 3 M HCl. Then
      add 10 drops of 0.2 M $BaCl_2$. The formation of a white precipitate under these conditions is
      the positive test for sulfate. The reaction is

$$SO_4^{2-} + BaCl_2 \rightarrow BaSO_4 + Cl^-$$

Test for chloride, Cl⁻

4. Place one drop of the solution labeled "Cl⁻" in a test tube. Add 20 drops of water, 2 drops of 3 M HNO_3 and shake. Now add 2 drops of 0.1 M $AgNO_3$. The white precipitate that forms is AgCl. The reaction is

$$AgNO_3 + Cl^- \rightarrow AgCl + NO_3^-$$

    In a positive test for chloride, this precipitate dissolves in strong ammonia solution and is then regenerated upon neutralizing the ammonia with the HNO_3. Perform Step 5 to observe this.

5. Centrifuge the solution from Step 4 for 2 minutes. Remove and discard the supernatant (the clear liquid above the precipitate) with a disposable pipet. Add 20 drops of water to wash the precipitate; shake, centrifuge, and again remove and discard the supernatant. Now add 8 drops of 6 M ammonia solution and shake. Did the precipitate dissolve completely? Add 20 drops of the 3 M HNO_3 and shake. Was the precipitate regenerated? Note that the precipitate is pure white.

6. Perform Steps 4 and 5 again twice, but with Br⁻ and I⁻ instead of Cl⁻. Precipitates form in these cases, too. The reactions are

$$AgNO_3 + Br^- \rightarrow AgBr + NO_3^-$$

and

$$AgNO_3 + I^- \rightarrow AgI + NO_3^-$$

    Note any differences in the color of the precipitates compared to AgCl and in the solubility in the ammonia solution.

Tests for bromide and iodide ions, Br and I

Note: For these steps, you will need fresh chlorine water. Chlorine water is prepared by bubbling chlorine gas through distilled water for several minutes. Your instructor may choose to prepare this for you ahead of time.

7. Place 4 drops of the solution labeled Br⁻ in a test tube. Add 20 drops of water and 5 drops of hexane and shake vigorously. The hexane does not dissolve in the water and forms a separate layer at the top of the test tube. Now add five drops of the chlorine water (more than five drops are needed if the chlorine water is weak) and shake again. The chlorine reacts with bromide in the water layer to give bromine, Br_2, which is more soluble in the hexane than in the water. It turns the hexane layer an orange color.

8. Repeat Step 7, but with the solution labeled I⁻ rather than Br⁻. Chlorine reacts with iodide to give iodine, I_2, which also is more soluble in the hexane, but this time the color is purple. The reactions of these ions with chlorine are

$$Br^- + Cl_2 \rightarrow Br_2 + Cl^-$$

and

$$I^- + Cl_2 \rightarrow I_2 + Cl^-$$

Test for nitrate ions, NO_3^-

Note: Be aware that Fe^{3+} and iodide give the same results as nitrate in this test. Thus, if you've determined that Fe^{3+} is present in your unknown, the presence of nitrate will need to be determined by the process of elimination.

9.   Perform this test with the help of your supervisor. Place 1 drop of the solution labeled "NO_3^-" in a cuvette. Add 1 drop of 0.1 M HCl and fill to the top with distilled water. Shake. Nitrate strongly absorbs ultraviolet light at 220 nanometers (nm) under these conditions. Wipe the exterior of the cuvette with a paper towel, and place it in an ultraviolet (UV) spectrophotometer set at 220 nm. A high absorbance reading represents a positive test for nitrate. Compare with distilled water.

Part C. Analysis of Unknowns

General approach

1.   The tests that you learned for the various ions may be run in any order. Keep in mind, however, that Fe^{3+}, I^-, and NO_3^- all give the same result in the UV absorption test.

2.   Check to see if any of the solutions have a color. If any do, make a guess as to what ion may be the cause of the color and run the test for that ion first. (Bear in mind that your supervisor may have used food coloring to give the solution the color it has, so the tests for the colored ions will need to be run.)

3.   Remember that each vial has just one kind of cation and one kind of anion. As soon as you've identified a cation in a given vial, you don't need to test for any other cation in that vial. The same is true of the anions.

4.   Run the easy tests first. It may only take a few minutes to complete your analysis of a given vial.

Format for Your Lab Notebook

Part A

Step 1. What is the color of a solution of Fe^{3+} ions? _____

Step 2. What is the color of a solution of Ni^{2+} ions? _____

Step 5. Describe the flame colors:

      a. Na^+: _____

      b. K^+: _____

      c. Ca^{2+}: _____

      d. Li^+: _____

Part B

Step 6. Describe the color of AgBr:

    Describe the solubility of AgBr in strong ammonia water: _____

    Describe the color of AgI: _____

    Describe the solubility of AgI in strong ammonia water: _____

Part C

Be sure to keep a clear accounting of your procedure for the unknowns in your laboratory notebook. Record your observations for each test run for each ion in each vial. Remember, there is only one kind of cation and one kind of anion in each. As soon as you have identified a cation, no further tests are needed for other cations. The same is true for the anions. You should determine for yourself which tests are the "easy" tests and run those first.

The following is a suggested format for recording the observations for each step performed.

Vial Number	**Test**	**Observation**	**Conclusion**

The "Results" section of your notebook writeup can take the following form:

Vial label _____   Cation found _____   Anion found _____

Vial label _____   Cation found _____   Anion found _____

Vial label _____   Cation found _____   Anion found _____

Vial label _____   Cation found _____   Anion found _____

Vial label _____   Cation found _____   Anion found _____

Reference

Kenkel, J., Kelter, P., and Hage D., *Chemistry: An Industry-Based Introduction,* CRC Press/Lewis Publishers, Boca Raton, FL, 2000, Chap. 1, Sec. 1.2, 1.8.

Experiment 4
Practice in Naming and Formula Writing

Introduction

This is an exercise in writing formulas and naming of inorganic compounds. Complete the following tables. You may use any written material available to assist. The first one is done for you. Your supervisor may ask you to read your table to another technician in the lab to practice pronouncing the names.

TABLE 3.1

Anion	Sodium		Magnesium	
	Formula	Name	Formula	Name
Fluoride	NaF	Sodium fluoride		
Cl^-				
Bromide				
I^-				
Oxide				
S^{2-}				
Hydride				
C^{4-}				
Nitrate				
SO_4^{2-}				
Hydroxide				
CO_3^{2-}				
Phosphate				
NO_2^-				
Sulfite				
PO_3^{3-}				
Hypochlorite				
BrO_2^-				
Iodate				
ClO_4^-				
Perbromate				
IO^-				
Bicarbonate				
HSO_4				
Bisulfite				
HPO_4^{2-}				
Dihydrogen phosphate				
CrO_4^{2-}				
Dichromate				
MnO_4^-				
Cyanide				
SCN^-				
Acetate				
$C_2O_4^{2-}$				
Thiosulfate				
Phosphate tribasic				
Phosphate dibasic				
Phosphate monobasic				

TABLE 3.2

Anion	Aluminum Formula	Aluminum Name	Ammonium Formula	Ammonium Name
Fluoride				
Cl^-				
Bromide				
I^-				
Oxide				
S^{2-}				
Hydride				
C^{4-}				
Nitrate				
SO_4^{2-}				
Hydroxide				
CO_3^{2-}				
Phosphate				
NO_2^-				
Sulfite				
PO_3^{3-}				
Hypochlorite				
BrO_2^-				
Iodate				
ClO_4^-				
Perbromate				
IO^-				
Bicarbonate				
HSO_4^-				
Bisulfite				
HPO_4^{2-}				
Dihydrogen phosphate				
CrO_4^{2-}				
Dichromate				
MnO_4^-				
Cyanide				
SCN^-				
Acetate				
$C_2O_4^{2-}$				
Thiosulfate				
Phosphate tribasic				
Phosphate dibasic				
Phosphate monobasic				

TABLE 3.3

	Copper (I)		Copper (II)	
Anion	Formula	Name	Formula	Name
Fluoride				
Cl^-				
Bromide				
I^-				
Oxide				
S^{2-}				
Hydride				
C^{4-}				
Nitrate				
SO_4^{2-}				
Hydroxide				
CO_3^{2-}				
Phosphate				
NO_2^-				
Sulfite				
PO_3^{3-}				
Hypochlorite				
BrO_2^-				
Iodate				
ClO_4^-				
Perbromate				
IO^-				
Bicarbonate				
HSO_4^-				
Bisulfite				
HPO_4^{2-}				
Dihydrogen phosphate				
CrO_4^{2-}				
Dichromate				
MnO_4^-				
Cyanide				
SCN^-				
Acetate				
$C_2O_4^{2-}$				
Thiosulfate				
Phosphate tribasic				
Phosphate dibasic				
Phosphate monobasic				

TABLE 3.4

Anion	Ferrous		Ferric	
	Formula	Name	Formula	Name
Fluoride				
Cl^-				
Bromide				
I^-				
Oxide				
S^{2-}				
Hydride				
C^{4-}				
Nitrate				
SO_4^{2-}				
Hydroxide				
CO_3^{2-}				
Phosphate				
NO_2^-				
Sulfite				
PO_3^{3-}				
Hypochlorite				
BrO_2^-				
Iodate				
ClO_4^-				
Perbromate				
IO^-				
Bicarbonate				
HSO_4^-				
Bisulfite				
HPO_4^{2-}				
Dihydrogen phosphate				
CrO_4^{2-}				
Dichromate				
MnO_4^-				
Cyanide				
SCN^-				
Acetate				
$C_2O_4^{2-}$				
Thiosulfate				
Phosphate tribasic				
Phosphate dibasic				
Phosphate monobasic				

TABLE 3.5

Anion	Stannous		Tin (IV)	
	Formula	Name	Formula	Name
Fluoride				
Cl^-				
Bromide				
I^-				
Oxide				
S^{2-}				
Hydride				
C^{4-}				
Nitrate				
SO_4^{2-}				
Hydroxide				
CO_3^{2-}				
Phosphate				
NO_2^-				
Sulfite				
PO_3^{3-}				
Hypochlorite				
BrO_2^-				
Iodate				
ClO_4^-				
Perbromate				
IO^-				
Bicarbonate				
HSO_4^-				
Bisulfite				
HPO_4^{2-}				
Dihydrogen phosphate				
CrO_4^{2-}				
Dichromate				
MnO_4^-				
Cyanide				
SCN^-				
Acetate				
$C_2O_4^{2-}$				
Thiosulfate				
Phosphate tribasic				
Phosphate dibasic				
Phosphate monobasic				

Reference

Kenkel J., Kelter P., and Hage D., *Chemistry: An Industry-Based Introduction,* CRC Press/Lewis Publishers, Boca Raton, FL, 2000, Chap. 2 and 4.

Experiment 5
A Chemical
Scavenger Hunt

I.O.N.S.
Innovative Options and New Solutions

Interoffice Memo From Claire Hemistry, CEO

It has come to my attention recently that many of our laboratory workers have not had much of an opportunity to explore the various information sources that are at our disposal. These include Material Safety Data Sheets (MSDS), the *Merck Index* on compact disc, and the *Handbook of Chemistry and Physics,* for example. In addition, some of our newer employees need to come up-to-speed on our system for ordering chemicals.

Therefore, I am requesting that all of our laboratory workers obtain a copy of the activity I have written called "A Chemical Scavenger Hunt" and try to complete it in the next week.

Since we are a consulting firm, many of our clients come to us with problems relating to consumer products that can be found in and around the home, including toothpaste, shampoo, food products, and the like. The scavenger hunt activity identifies a number of chemicals that we often have needed in our stock and that can be found in a number of these consumer products. It then asks you to find a consumer product in which each chemical is found and investigate the various information sources and find the answer to some questions about these chemicals. You will notice that the activity calls for you to look up the MSDS sheet and find certain specific information. This also gives you valuable experience with the MSDS sheet. You may use the MSDS sheets found in the lab or look them up on the Internet.

Good luck, and please inform your supervisor when you have completed the activity so that it can be documented.

C. Hemistry
Claire Hemistry, CEO

Introduction

Many of the simple chemicals we encounter in our chemistry studies can be found on the ingredients lists on the containers of consumer products such as foods, beverages, pharmaceuticals, personal hygiene products, household cleansers, fertilizers, and various other items we buy in grocery stores, pharmacies, hardware stores, and the like. Information about these chemicals, such as properties, safety hazards, formulas and formula weights, and purchasing information, can be found in a wide variety of reference sources, such as handbooks, computer data banks, the Internet, chemical catalogs, and the labels on laboratory containers. In this laboratory exercise, you are given a list of 13 chemicals (see the Data sheet) and are asked to examine the ingredients lists on various consumer products and the reference sources listed above in order to determine the consumer product in which each chemical is found and to also record certain other information about each. The consumer products are shampoo, beef jerky, English muffins, toothpaste, baking soda, boxed macaroni and cheese, epsom salt, iodized salt, washing soda, liquid drain opener, nasal decongestant, baking powder, and ice melt.

The reference sources to be used are the *Handbook of Chemistry and Physics* (published by CRC Press LLC); the *Merck Index CD-ROM* (published by Chapman and Hall); the Material Safety Data Sheet (MSDS) found in the lab or on the Internet, such as through the Fisher Chemical Company Web site (www.fisher1.com), or others suggested by your supervisor; the labels on the stock containers of the chemicals found in the laboratory; and the Fisher Chemical Company chemicals catalog. Your supervisor will demonstrate the use of each of these. Your supervisor may or may not require you to use your laboratory notebook for this work. Accordingly, a data sheet is provided.

Check with your supervisor to see if you should do all 13 chemicals.

DATA SHEET   Name _____

1. Sodium Nitrite

Consumer Product: _____

Formula: _____

From the *Handbook of Chemistry and Physics:* Melting Point: _____

From the *Merck Index CD-ROM:*

> What is the formula of the brown fumes evolved when it is decomposed with weak acids? _____

From the MSDS on the Internet:

> For fighting fires near it, what should be the extinguishing media? _____

From the label on the container of this chemical: F.W.: _____

From the Fisher Chemical Catalog:

> The cost of 250 g of certified ACS grade chemical:_____
> The catalog number for the above: _____

2. Ammonium Sulfate

Consumer Product: _____

Formula: _____

From the *Handbook of Chemistry and Physics:* Density: _____

From the *Merck Index CD-ROM:*

> What two adjectives are used to describe its crystals?
>
> _____
> _____

From the MSDS on the Internet:

> For disposal, is it listed as a material banned
> from land disposal according to the RCRA? _____

From the label on the container of this chemical: F.W.: _____

From the Fisher Chemical Catalog: _____

> The cost of 500 g of certified ACS grade chemical:_____
> The catalog number for the above: _____

3. Sodium Fluoride

Consumer Product: _____

Formula: _____

From the *Handbook of Chemistry and Physics:* Boiling Point: _____

From the *Merck Index CD-ROM:*

> Can it be used to fluoridate drinking water?_____

From the MSDS on the Internet:

> Under Stability and Reactivity, what is listed as "conditions to avoid?"
>
> _____

From the label on the container of this chemical: F.W.: _____

From the Fisher Chemical Catalog:

> The cost of 100 g of certified ACS grade chemical:_____
> The catalog number for the above: _____

4. Sodium Bicarbonate

Consumer Product: _____

Formula: _____

From the *Handbook of Chemistry and Physics* (Hint: Look under sodium carbonate hydrogen):

> Density:_____

From the *Merck Index CD-ROM:*

> How hot do you have to heat it to convert it to Na_2CO_3? _____

From the MSDS on the Internet:

> What eye protection is required?_____

From the label on the container of this chemical: F.W.: _____

From the Fisher Chemical Catalog:

> The cost of 500 g of certified ACS grade chemical:_____
> The catalog number for the above: _____

5. Potassium Chloride

Consumer Product: _____

Formula: _____

From the *Handbook of Chemistry and Physics:* Synonym: _____

From the *Merck Index CD-ROM:*

> How many mL of water will 1 g of it dissolve in? _____

From the MSDS on the Internet:

> What does IATA say about transporting this chemical?_____

From the label on the container of this chemical: F.W.: _____

From the Fisher Chemical Catalog:

> The cost of 500 g of certified ACS grade chemical:_____
> The catalog number for the above: _____

6. Sodium Hypochlorite

Consumer Product: _____

Formula: _____

From the *Handbook of Chemistry and Physics* (Hint: Look under sodium chlorite):

Crystalline form:_____

From the *Merck Index CD-ROM:*

What are three names given to aqueous solutions of sodium hypochlorite?

_____

_____

_____

From the MSDS on the Internet:

Under Section 9, what odor does it have?_____

From the label on the container of this chemical: F.W.: _____

From the Fisher Chemical Catalog:

The cost of 1 l of 5.65 to 6% solution:_____
The catalog number for the above: _____

7. Calcium Phosphate, Dibasic

Consumer Product: _____

Formula: _____

From the *Handbook of Chemistry and Physics* (Hint: Look for calcium orthophoshate, di-(sec)):

Density:_____

From the *Merck Index CD-ROM:*

Why is it placed in foods?_____

From the MSDS on the Internet:

Under Section 3, what does it say about the effects on the eyes? _____

_____

From the label on the container of this chemical: F.W.: _____

From the Fisher Chemical Catalog:

The cost of 500 g of certified grade chemical: _____
The catalog number for the above: _____

8. Calcium Sulfate

Consumer Product: _____

Formula: _____

From the *Handbook of Chemistry and Physics* (look under "soluble anhydrite"):

Density:_____

From the *Merck Index CD-ROM:*

> By what name is it known as a desiccant used in laboratory and industry?

_____

From the MSDS on the Internet:

> Under Section 9, what is its appearance (for calcium sulfate hemihydrate)?

_____

From the label on the container of this chemical: F.W.: _____

From the Fisher Chemical Catalog:

> The cost of 10 kg of calcium sulfate hemihydrate: _____
> The catalog number for the above: _____

9. Ferrous Sulfate

Consumer Product: _____

Formula: _____

From the *Handbook of Chemistry and Physics* (look under the IUPAC name):

> Density:_____

From the *Merck Index CD-ROM:*

> For what "test" is it used for in qualitative analysis?_____

From the MSDS on the Internet:

> For Iron (II) sulfate heptahydrate, under Section 10, what are some hazardous decomposition products? _____

From the label on the container of this chemical: F.W.: _____

From the Fisher Chemical Catalog:

> The cost of 500 g of the heptahydrate certified ACS grade chemical: _____
> The catalog number for the above: _____

10. Calcium Chloride

Consumer Product: _____

Formula: _____

From the *Handbook of Chemistry and Physics:* Melting Point: _____

From the *Merck Index CD-ROM:*

> What is it used for on unpaved roads? _____

_____

From the MSDS on the Internet:

> Under Section 15, according to the Clean Air Act,
> does it present an air pollution hazard?_____

From the label on the container of this chemical: F.W.: _____

From the Fisher Chemical Catalog:

        The cost of 500 g of the anhydrous certified grade chemical: _____

        The catalog number for the above: _____

11. Sodium Thiosulfate

Consumer Product: _____

Formula: _____

From the *Handbook of Chemistry and Physics:* Density: _____

From the *Merck Index CD-ROM:*

        Do the crystals have an odor? _____

From the MSDS on the Internet:

        Under Section 4, what do you give a person who has ingested it?

        _____

From the label on the container of this chemical: F.W.: _____

From the Fisher Chemical Catalog:

        The cost of 500 g of the anhydrous certified grade chemical: _____

        The catalog number for the above: _____

12. Sodium Carbonate

Consumer Product: _____

Formula: _____

From the *Handbook of Chemistry and Physics:* Melting Point: _____

From the *Merck Index CD-ROM:*

        What grade of this chemical is known as "soda ash?" _____

From the MSDS on the Internet:

        Under Section 5, does this material burn? _____

From the label on the container of this chemical: F.W.: _____

From the Fisher Chemical Catalog:

        The cost of 500 g of the anhydrous certified ACS grade chemical:_____

        The catalog number for the above: _____

13. Magnesium Sulfate

Consumer Product: _____

Formula: _____

From the *Handbook of Chemistry and Physics:* Density: _____

From the *Merck Index CD-ROM:*

        Occurs in nature as what mineral? _____

From the MSDS on the Internet:

>   Under Section 16, on what date was this MSDS created? _____

From the label on the container of this chemical: F.W.: _____

From the Fisher Chemical Catalog:

>   The cost of 500 g of the anhydrous certified grade chemical: _____
>   The catalog number for the above: _____

Note

For nomenclature and formula writing, see Kenkel J., Kelter P., and Hage D., *Chemistry: An Industry-Based Introduction,* CRC Press/Lewis Publishers, Boca Raton, FL, 2000, Chap. 2, 4.

Experiment 6
The Protective
and Emergency
Equipment Exercises

I.O.N.S.
Innovative Options and New Solutions

Interoffice Memo From Claire Hemistry, CEO

Dear I.O.N.S. Staff:

Since the safety of our laboratory personnel and visitors to the laboratory is always a primary concern, I would like all our laboratory workers to participate in this exercise in order to increase their awareness of the proper use and testing of our protective and emergency equipment while at the same time determining if this equipment is working properly. The items to be tested are the fume hoods, the eyewash stations, and the safety showers.

For the procedure, I would like to refer you to the a book entitled *Building Student Safety Habits for the Workplace,* developed recently by the Partnership for the Advancement of Chemical Technology (PACT), centered at Miami University/Middletown, Middletown, OH, with funding from the National Science Foundation. The specific exercises to be performed are Exercise 4E in which a fume hood is examined and tested, and Exercise 7C in which eyewash stations and safety showers are examined and tested.

Each laboratory worker should obtain a photocopy of the relevant pages in this book (those labeled "student page" in the Instructor Edition) and follow through with the examination and testing, answering the questions and filling in the tables. Please use your regular laboratory notebook for the recordkeeping and cut out the tables from the photocopied pages and tape them into your notebook.

When finished, please draft a memo to me stating that you have completed the exercises and noting any problems and recommendations with regard to the equipment tested.

Sincerely,

Claire
Claire Hemistry, CEO

Acknowledgment

Permission to link this experiment to the book *Building Student Safety Habits for the Workplace,* developed by the Partnership for the Advancement of Chemical Technology and published by Terrific Science Press, Miami University, Middletown, OH, is gratefully acknowledged.

Section **4**

I.O.N.S.
Client Projects

Innovative Options and New Solutions

Contents

Experiment 7
Out of Spec —
Out of Mind

I.O.N.S.
Innovative Options and New Solutions

Interoffice Memo From Claire Hemistry, CEO

Dear I.O.N.S. Staff:

We have acquired a contract from a company called The Solution Makers located in northern New Jersey. This company prepares and sells certified standard solutions for use in accredited laboratories worldwide. They are especially known for their high-quality atomic absorption standard solutions widely used to calibrate atomic spectroscopy instruments as well as other instruments. The chemicals they use to prepare these solutions are purchased from The Inorganic Chemical Company of North America (ICCNA).

Workers at The Solution Makers noticed that recent shipments of solid iron (III) chloride and solid nickel (II) chloride purchased from ICCNA to make 1000 ppm solutions of iron and nickel appeared to be a slightly different color compared to the same chemicals purchased earlier. When they prepared the solutions, they too appeared to be a slightly different color compared to solutions prepared earlier. They are suspicious that these chemicals are "out-of spec" (meaning they do not meet the specifications required by law) due to a contaminant. Before they start pointing fingers and say that ICCNA's chemicals are out-of-spec, however, they would like to have an independent analysis done by our laboratory. The contract does not call for us to identify a contaminant. Our purpose is to simply discover whether or not there is, in fact, a colored contaminant present. The Solution Makers have given us samples of both the clean iron (III) chloride and nickel (II) chloride and also the chemicals suspected of being contaminated.

Because of his reputable work with off-color and contaminated materials, I have asked Dr. Otto Speck of Colorama Consultants to give us some direction. His memo and suggested procedure are attached. We also have a memo from Professor Will B. Green of Shade University commenting on Dr. Speck's suggestions. Please address your report on this one to Dr. Mary Kay Downthyme of The Solution Makers.

Claire

olorama   *onsultants*

Dr. Otto Speck

Dear I.O.N.S. Staff:

Since your contract does not call for you to identify the contaminant, only determine whether one is present, I think I have a simple method for your project. Both the iron (III) chloride and the nickel (II) chloride have a color of their own. Their solutions also have a color of their own. Iron (III) chloride has a yellow color when dissolved in water and nickel (II) chloride has a green color. This indicates that they both absorb wavelengths of light in the visible region of the spectrum. If they didn't, then there wouldn't be color at all.

I would suggest that you use your visible spectrophotometer, sometimes called a colorim-eter, to determine the pattern of absorption of light in the visible region, i.e., to obtain the visible transmittance or absorption spectra of the solutions. Both iron(III) chloride and nickel (II) chloride have their own unique pattern of absorption. You could then also obtain the visible transmittance or absorption spectra of the solutions of the off-color samples. If there are significant differences in the patterns of the pure and the off-color, then there must be a contaminant.

The attached procedure calls for you to first prepare solutions of all four samples. Then, using your visible spectrophotometer, you will find the % transmittance or absorbance reading for a series of wavelengths in the visible region for each sample. If you plot the % transmittance (or absorbance) on the y-axis and the wavelength on the x-axis, you will have the visible transmittance (or absorption) spectra of all four materials and you can visually compare the patterns. You will then know whether there is a contaminant.

Good Luck!

Otto Speck

Shade University

From Dr. Will B. Green

Dr. Speck suggests that you look for significant differences in the pattern of absorption. The pattern is the shape of the transmittance (or absorbance) "curve" that your plot will have. Usually these curves will display fairly smooth peaks and valleys because the different wavelengths are absorbed to different degrees. If there is a contaminant that absorbs more of a particular wavelength (or range of wavelengths) than the pure material, then the shapes of these curves will be altered. The alteration could take the form of a "bump" that wasn't there for the pure material, or a change in a wider range of wavelengths to higher or lower levels. One caution, however, is that if the entire curve of the suspect material is raised or lowered (rather than its shape altered), it is likely due to a concentration difference and not a contaminant.

One additional point is that it is possible for a contaminant to not have a color or to not absorb in the visible region of the spectrum. In that case, the color of the material and their solutions may still be visibly different but it will not show up in the spectra. You should make this point to your client in the event your results do not show a contaminant. You also could then tell the client that you would seek another contract with them to look for a nonabsorbing contaminant.

Will

I.O.N.S. Safety Report

prepared by Ben Whell, I.O.N.S. Safety Coordinator

Experiment 7

Equipment and Technique

- Avoid spilling solutions on spectrophotometer (potential electrical safety hazard).
- Wear gloves to avoid contact with chemicals and solutions.
- Do not use a spectrophotometer that has a frayed cord.

Chemicals

- Avoid contact with the contaminated samples, especially since you do not know what the contaminant is.

Workplace Cleanup

- Consult your supervisor for advice on disposal of an unidentified contaminant.
- Rinse beakers with water and dry with a towel before storing.
- If cuvettes are not of the disposable variety, rinse with distilled water and dry outside with a towel before storing. Store inverted in rack.

Hazards Classifications

- Potential injuries are minor and treatable on site.

Laboratory Safety Quiz

1. List some potential safety hazards a lab worker may encounter when a contaminant has not been identified. How should the worker approach such a situation. What about disposal?

 *olorama*    *onsultants*

Dr. Otto Speck

Procedure for Measuring the Visible Transmittance Spectrum of a Solution

1. Turn on the visible spectrophotometer and allow it to warm up for 15 minutes.

2. Note the color of the four samples, comparing the colors of the pure with the off-color for both the iron (III) chloride and the nickel (II) chloride. Write your descriptions in your laboratory notebook. Prepare solutions of each of the four samples by weighing 0.5 g of each into a 50-mL beaker and adding 30 mL of water. Stir thoroughly until dissolved. Label each beaker. Write descriptions of the color of each of the solutions in your notebook.

3. Clean five cuvettes used with your spectrophotometer (inside and outside), but DO NOT SCRATCH THEM (use a cotton swab with soapy water). Rinse thoroughly with distilled water.

4. Prepare the cuvettes by rinsing each with the particular solution that will be placed in it. Fill four cuvettes with the four solutions, labeling each with a small label near the top. Fill the fifth with distilled water. This latter cuvette will be used as a "blank."

5. Set the spectrophotometer to 750 nanometers (nm). Ask your supervisor to show you how to calibrate the spectrophotometer with the blank, since there can be differences from one instrument to another. Then, measure the %T for all four solutions.

6. Repeat Step 5 but at a wavelength of 730 nm, then 710 nm, etc., recording a reading every 20 nm until you reach 450 nm.

7. When finished, you should have a total of 64 readings, 16 for each solution covering wavelengths in the visible region from 750 nm to 450 nm. Plot these as suggested in my memo, all four on the same graph.

8. Compare the spectra of the "clean" samples to the spectra for the suspect samples with regard to the pattern or shape of the curves. The patterns should be identical if there is no contaminant.

References

Kenkel J., Kelter P., and Hage D., *Chemistry: An Industry-Based Introduction*, CRC Press/Lewis Publishers, Boca Raton, FL, 2000, Chap. 3.

Experiment 8
Assuring the Quality of a Copper Reference Standard

I.O.N.S.
Innovative Options and New Solutions

Interoffice Memo From Claire Hemistry, CEO

Dear I.O.N.S. Staff:

Dr. Mary Kay Downthyme of the The Solution Makers was so pleased with our work on the contaminated chemicals from ICCNA (Experiment 7) that she has come back to I.O.N.S. with another project. Since their main business is to prepare certified reference standard solutions for their customers, their laboratory workers spend much time running tests on these solutions to assure their quality. For example, the atomic absorption standard solutions that were earlier suspected by their lab as being contaminated must have a concentration of the indicated metal of 1000 ppm. Their quality assurance laboratory runs quality checks daily on these solutions and the solutions cannot be sold until the laboratory issues the certificate that states that they indeed have the required concentration. In addition, an expiration date must be stamped on the labels of these solutions and so, occasionally, they run tests on solutions that have been manufactured previously but not shipped.

In order to assure that their laboratory is obtaining the correct results on these solutions, they will periodically contract with an outside laboratory to do a check of their work. I.O.N.S. has been asked to help in this regard. We have a sample from them. It is a "blind" sample in that we do not know the history, i.e., we do not know if it is a regular quality assurance sample, whether it is one for which the expiration date has passed, or whether it is one for which they have received a customer complaint. Thus, we cannot make any judgment about the sample — only report the results. It is labeled as a 1000 ppm copper solution.

I have asked Professor Q. Wally Tee of the State University of Metals as well as Ralph Blue of the Federal Institute for Standards and Technology (FIST) to consult on this project and their memos are attached. Please report the results of your work to Dr. Downthyme at The Solution Makers, Inc.

Claire

C. Hemistry

State University of Metals

Providing the Keys to Success in Life

Claire:

Thanks for asking me to consult on this interesting project. There is a nice "colorimetric" method of analysis by which the copper present in water solution can be determined. As you know, solutions containing copper have a blue color. This blue color can be greatly enhanced by the addition of ammonia (or ammonium hydroxide) to the solution. This enhanced blue color is more accurately measured by a spectrophotometer. The procedure that I suggest measures this enhanced blue color and relates it to how much copper is present. The more copper, the more color. The amount of color can be determined by measuring the absorbance with a spectrophotometer that operates in the visible region of the electromagnetic spectrum.

If your technicians will prepare a series of standard solutions representing different masses of copper (measured in milligrams) and measure the absorbances of each of these standards, one can make a graph of absorbance vs. milligrams. Such a graph, called a "standard curve," should be linear in the range that I suggested in the attached procedure. One can then determine the concentration of the submitted sample by measuring its absorbance and then looking at the standard curve to determine the concentration that corresponds to the measured absorbance of this sample.

Good Luck!

Wally

Professor Q. Wally Tee

F.I.S.T.

FEDERAL INSTITUTE OF STANDARDS AND TECHNOLOGY

Ms. Hemistry:

I am pleased that you contacted me regarding this project. As you know, all certified reference materials (CRMs) sold in this country must be traceable to the Federal Institute of Standards and Technology (FIST). In other words, the solutions manufactured by The Solutions Makers must have been tested using standard reference materials (SRMs) formulated by FIST. I have read the procedure suggested by Professor Tee. The copper sulfate required is available as a SRM from FIST and I will forward the catalog information to you.

There have been no previously-reported problems with the products manufactured by The Solution Makers. I assume that the reason for your involvement is a routine external quality check. Keep in mind that the 1000 ppm solutions have a concentration tolerance of 10 ppm in order to comply with federal regulations. This means that results between 990 ppm and 1010 ppm are acceptable for the solution you are testing.

If you have any further questions, please contact me.

Sincerely,

Ralph Blue

Ralph Blue
F.I.S.T.

I.O.N.S. Safety Report
prepared by Ben Whell, I.O.N.S. Safety Coordinator

Experiment 8

Equipment and Technique

- Use plastic dropper bottle for 6 M ammonium hydroxide.
- Consider using a fume hood to vent ammonia fumes.
- Avoid spilling solutions on spectrophotometer (potential electrical safety hazard).
- Wear gloves to avoid contact with chemicals and solutions.
- Do not use a spectrophotometer that has a frayed cord.

Chemicals

- The 6 M ammonium hydroxide has a very strong ammonia odor. Avoid breathing the vapors.

Workplace Cleanup

- Pour copper solutions into a designated waste container. Do not discharge down the drain.
- Rinse beakers with water and dry with a towel before storing.
- If cuvettes are not of the disposable variety, rinse with distilled water and dry outside with a towel before storing. Store inverted in rack.

Hazards Classifications

- Potential injuries are minor and treatable on site.

Laboratory Safety Quiz

1. Why should one be cautious when attempting to detect the odor of a solution? Describe what may happen if the odor is overwhelmingly strong.

State University of Metals

Providing the Keys to Success in Life

Procedure for the Analysis of a Copper
Reference Standard by the I.O.N.S. Corporation

by Professor Q. Wally Tee

1. Weigh 0.982 g of $CuSO_4 \cdot 5H_2O$ on a piece of weighing paper and transfer to a clean 100-mLmL volumetric flask using a funnel. Add water to rinse the funnel and half fill the flask and swirl until the solid is dissolved. Then add more water so that the bottom of the meniscus rests on the 100-mL line. Place the stopper in the flask and thoroughly mix this solution to make it homogeneous. This solution has 250 mg of copper in it (2.50 mg/mL).

2. Obtain six clean 10-mL graduated cylinders. Place a small piece of labeling tape on the base of each. To one, add precisely 5.00 mL of the copper solution prepared in Step 1. Write the number "5" on the label. To another, add precisely 3.00 mL of the copper solution and write the number "3" on the label. To a third, add precisely 1.00 mL of the copper solution and write the number "1" on the label. To a fourth, add precisely 0.50 mL of the copper solution and write the number "0.5" on the label. To a fifth, add precisely 5.00 mL of the 1000-ppm solution you are testing. Write "test" on the label to indicate that it is the copper solution you are testing. To the last graduated cylinder, add nothing and write "blank" on the label to indicate no copper present.

3. Add distilled water to each graduated cylinder so that the meniscus lies on the 8-mL line. Then add 6 M ammonium hydroxide solution (*caution:* strong odor; use a plastic dropper bottle) to each so that the meniscus lies on the 10-mL line. Now shake or stir all solutions by moving a stirring rod (flattened on one end) up and down in the cylinder a number of times being very careful not to contaminate them (such as with a wet stirring rod). All solutions, except for the blank, should now be a blue color that gets darker as the copper content increases. The test solution should display a medium blue color. The blank solution should be completely colorless.

4. In the Data section of your notebook, create a table with three columns. The first column should be headed "mL" to indicate the number of mL of the copper solution present in each (5.00, 3.00, etc.). The second column should be headed "mg" to indicate the number of milligrams of copper present in each. The solution with 5.00 mL in it has 12.5 mg of copper (since there are 2.50 mg in each mL, as was determined in Step 1), the solution with 3.00 mL has 7.50 mg in it, the solution with 1.00 mL has 2.50 mg in it, and the solution with 0.50 mL in it has 1.25 mg of copper in it.

5. Now use a spectrophotometer to measure the absorbance of each solution at a wavelength of 610 nm. This is the wavelength of visible light that is absorbed the most by these solutions. The blank is used to calibrate. Record the absorbance in the third column (headed ABS) of the data table. Consult with your supervisor if you need assistance with the operation of the spectrophotometer or with any calculations.

6. This step may be done with a computer. Create a graph with absorbance on the y-axis and milligrams on the x-axis and draw the best straight line you can through the points you have plotted. Determine the number of milligrams of copper in the test solution from this graph. To calculate the ppm of copper in the original undiluted test solution, multiply the number of milligrams found from the graph by 200. This is done because ppm corresponds to milligrams per liter (1000 mL) and one liter is a volume that is 200 times larger than the 5.00 mL used in Step 2.

References

Kenkel J., Kelter P., and Hage D., *Chemistry: An Industry-Based Introduction*, CRC Press/Lewis Publishers, Boca Raton, FL, 2000, Chap. 3.

Experiment 9
No Labels?
Now What?

I.O.N.S.
Innovative Options and New Solutions

Interoffice Memo From Claire Hemistry, CEO

Dear I.O.N.S. Staff:

The Petroleum Chemical Company (Petrochemco, Inc.) has come to us with an interesting problem. Recently in their plant, an inexperienced laboratory technician was sent from the lab into the plant to collect samples of the various liquids that their company manufactures. The problem is that he forgot to take labeling materials with him. When he began to take samples, it quickly became apparent that he would need to find some means to distinguish one sample from another. It was a long walk back to the lab, so he decided to merely place the sample bottles into a box in the order in which he took them, left to right. He was sure he would remember his sampling order and then he could label them when he returned to the lab.

Well, he did not remember the exact order when he returned, so he had 10 bottles and did not know for certain what was inside any of them. At this point, resampling is not an option because of the time period that has passed. Petrochemco has come to us for the purpose of establishing the identity of each liquid.

I have enlisted the help of Dr. Mista Ree of Pandamony University as our academic consultant and Calvin Viscus of Petrochemco for our industrial consultant on this project. Their attached memos give particularly good insight into how we should proceed.

When the work is completed, please write a memorandum to Calvin at Petrochemco, Inc., presenting him with the results of your work.

C. Hemistry

Claire Hemistry

Petrochemco, Inc.

From the desk of Calvin Viscus

In this project, distinguishing properties of the 10 organic liquids should be observed (Part A) and unknowns subsequently identified (Part B) according to an SOP which I wrote for this. The properties are (1) water miscibility, (2) density, (3) viscosity, (4) refractive index, and (5) odor. The 10 organic liquids are acetone, methanol, ethanol, isopropyl alcohol, heptane, cyclohexane, toluene, methyl ethyl ketone, butanol, and ethyl acetate.

For Part A, you should work as part of a team. Five teams will be needed. Each team should rotate through four stations in the lab, one for the water miscibility test, one for density measurement, one for viscosity measurement, and one for refractive index measurement. Each team should be assigned 2 of the 10 liquids to test at each station. As the results are obtained at each station, each team should record the results on the "grid" similar to the example given in the SOP. Perhaps you could put the same grid on a blackboard accessible and visible to each team. At the end of Part A, each worker can then fill in the entire "grid" in his/her notebook and have a complete set of data for the 10 compounds. This data will then be used in Part B to identify unknowns that are issued to each worker. In Part B, each worker should test several unknowns and identify each based on how well their properties match up with those measured in Part A. The property of odor may be used in Part B for confirmation if desired, comparing unknowns with knowns. In Part B, it may not be necessary to measure all five properties if you can identify your unknowns with just a few tests.

Good luck, and I will be anxious to get the results of your work.

Cal

Pandamony University

Professor Mista Ree

Dear I.O.N.S Lab Staff:

Some liquid substances are able to mix completely with water while others will mix only partially or not at all. If a substance mixes freely with water, only one liquid layer will be seen in the container. If a substance does not mix freely with water, two separate liquid layers will be observed in the container. With the water miscibility test, we observe this ability or lack of ability to mix. If an unknown mixes freely with water, we can conclude with certainty that it is not the same liquid as any liquid studied in Part A that does not mix with water. While at the same time we can conclude that it may be one of those that does mix with water. Thus, this test helps to narrow down the possibilities of what an unknown's identity might be.

Density is defined as the mass, or weight, a substance has per unit volume. Density is a physical characteristic of the substance that can be used to help identify it. If we measure the weight of a known volume of the substance and divide this weight by the volume, we would have the substance's density and if we compare this with density data for a set of known substances, it may help identify it.

Viscosity is a measure of a liquid substance's resistance to flow. Some liquids have a low apparent resistance to flow, while others are able to flow only very slowly and, thus, have a high resistance to flow. For example, maple syrup has a higher viscosity than does water. Viscosity is a characteristic of all liquid substances and can help identify them. It can be measured by measuring the time it takes for a known volume of liquid to pass vertically through a constriction. By measuring viscosities of unknowns and comparing them to viscosities of some knowns, the identities of the unknowns may be established.

Another property that can aid in the identification of liquid unknowns is refractive index. Refractive index is a measure of the "bending" of light through the liquid and can be measured with a device called a refractometer. Light travels through different liquids at different rates and, thus, each liquid appears to bend light at different angles.

Misty

I.O.N.S. Safety Report
prepared by Ben Whell, I.O.N.S. Safety Coordinator

Experiment 9

Equipment and Technique

- Be sure the laboratory is well ventilated; use a fume hood for all tests if there is sufficient space.
- Avoid smelling the liquids up close. The odor test should be done last and only if the liquid has not been identified with certainty.
- Keep the vessels containing the liquids capped at all times except when needed.
- Wear latex gloves at all times when handling these liquids.
- Do not remove safety glasses for any reason while in the laboratory.
- Use a rack that test tubes do not slip out of easily.
- Do not use a refractometer with a frayed cord.

Chemicals

- Handle liquids with care. Use droppers to transfer from original container to test tubes, graduated cylinders, etc.
- The acute toxicity of the chemicals tested range from slight to low. However, prudent precautions should be taken to avoid excessive inhalation and contact. Under no circumstance should any of these liquids be ingested.
- Most of the liquids tested are highly flammable. The use of open flames is strictly prohibited.
- Use a "wafting" technique to observe an odor. Ask your supervisor to demonstrate.
- Consult MSDS sheets for specific handling issues and questions.

Workplace Cleanup

- Dispose of liquids and rinsings only as directed by your supervisor.

Hazards Classifications

- Potential injuries are minor and treatable on site. If fumes cause lightheadedness, seek fresh air immediately.

Laboratory Safety Quiz

1. What should you do and not do if your safety glasses get in the way of your reading the refractometer?
2. What should you do if the person next to you spills a small container of a liquid on your bench top?
3. What should you do if you become dizzy due to the odors of the liquids?

Petrochemco, Inc.

SOP #98e

Procedure

Part A. Observation and Measurement of Known Liquids

All glassware used should be dry in order to avoid contamination that would give erroneous results. If you are using glassware that other lab workers used and it is wet, it can be first rinsed with acetone and the acetone either allowed to evaporate or rinsed out with the liquid to be tested.

1. Test 1: Water Miscibility

Place about 3 mL of water in a small test tube and add 20 drops of the liquid to be tested. Agitate the tube for about a minute, then place it in a test tube rack and wait for another minute. Observe the liquid in the test tube. If some of the organic liquid is still present as a separate phase, it is considered immiscible in water at room temperature. If you are unsure of the results, add a few more drops of the liquid to be tested and see if what you think may be the top layer gets larger. Placing a white or colored background behind the test tube also may help. Record your observations on the data grid, either miscible or immiscible.

Repeat with the other assigned liquid and record the results for both in the data grid and on the blackboard.

2. Test 2: Density

To measure density, the weight of a known volume of the liquid is measured and this weight is then divided by the volume to obtain density. For the volume measurement, you will use either a 10-mL buret or a 10-mL graduated cylinder. Make sure the buret or cylinder to be used is as dry as possible or rinsed in the manner described at the beginning of this SOP. (A buret is a graduated cylinder with a "stopcock" valve at the bottom.) Then, weigh a small (dry) beaker on the balance provided. Record this weight in your notebook.

If a buret is to be used, fill it to several milliliters above the 0.00 mL line and open the stopcock wide open to allow any air that will push through the tip to flow out. Stop the flow of the liquid when the bottom of the meniscus is either on the 0.00 mL graduation line or below it. If it is below, record the reading. Now, dispense the liquid to the beaker until the bottom of the meniscus is on or near the 10.00 mL line. Record the reading and subtract the first reading from the second. This is the volume of the liquid in the beaker (used for the density calculation).

If a graduated cylinder is used, fill it so that the bottom of the meniscus lies on or near (but under) the 10-mL line. Record this reading. This is the volume of the liquid used for the density calculation. Add the contents of the cylinder to the beaker. Measure the weight of the beaker again and record. Subtract the weight of the empty beaker from the weight of the beaker with the liquid in it. This is the weight of the liquid. Divide the weight of the liquid by the volume to obtain the density. Record on the data grid.

Repeat with the other assigned liquid and record the results on your data grid and also the grid on the blackboard.

3. Test 3: Viscosity

To measure viscosity, you will measure the time it takes for a certain volume of the liquid to pass through a constriction. A 50-mL buret that has been cut off at about the 38-mL line (and fire polished) can be used for this test. Make sure it is dry or rinsed with the liquid to be tested as directed at the beginning of this SOP. Using a stopwatch, or a watch with a second hand, measure the time it takes for 10 mL of the liquid tested to run out of the buret. Do this as carefully as you can so as to get an accurate time measurement for the exact volume used. Record this time on the data grid.

Repeat with the other assigned liquid and record the results on the data grid and also on the grid on the blackboard.

4. Test 4: Refractive Index

Your supervisor will show you how to use the refractometer. Measure the refractive index of both liquids assigned to you and record on the data grid and on the grid on the blackboard.

5. Conclusion

After all laboratory workers have made all measurements and all results are recorded on the blackboard, complete your data grid in your notebook with the data from the other groups.

Suggested Data Grid

Liquid	Water Miscibility	Density	Viscosity	Refractive Index
Acetone				
Methanol				
Ethanol				
Isopropyl alcohol				
Heptane				
Cyclohexane				
Toluene				
Methyl ethyl ketone				
Butanol				
Ethyl acetate				

Part B. Identification of Unknowns

1. Record the unknown numbers of your assigned unknowns.
2. Perform the water miscibility test first to narrow down the possibilities for your unknown. Then select and perform whatever test(s) will identify the unknowns. Be sure to keep good records in your notebook. When you think you have identified your assigned unknowns, compare the odors of the unknowns with odors of the pure liquids. This will serve as a final confirmation of your results.

Part C. Unknown Analysis

Suggested Format for Notebooks (For Each Unknown):

Unknown Number or Letter: _____

Results of Water Miscibility Test (miscible or immiscible): _____

        Possibilities for unknown identity based on miscibility:

        _____

        _____

Results of Density Test: _____

        Liquids with densities closest to unknown:

        Liquid name_____   Density_____

        Liquid name_____   Density_____

        Liquid name_____   Density_____

        Liquid name_____   Density_____

Results of Viscosity Test (seconds): _____

        Liquids with viscosities closest to unknown:

        Liquid name_____   Viscosity_____

        Liquid name_____   Viscosity_____

        Liquid name_____   Viscosity_____

        Liquid name_____   Viscosity_____

Results of Refractive Index Test: _____

        Liquids with refractive index closest to unknown:

        Liquid name_____   Refractive Index_____

Liquid name_____ Refractive Index_____

Liquid name_____ Refractive Index_____

Liquid name_____ Refractive Index_____

Results of Odor Test:

The odor of the unknown most closely matches that of the following liquids:

_____

_____

The unknown is _____

Reference

Kenkel J., Kelter P., and Hage D., *Chemistry: An Industry-Based Introduction*, CRC Press/Lewis Publishers, Boca Raton, FL, 2000, Chap. 6.

Experiment 10
The Frustrating
Federal Film Folly

I.O.N.S.
Innovative Options and New Solutions

Interoffice Memo From Claire Hemistry, CEO

Dear I.O.N.S. Staff:

I'm sure you are well aware of the recent nationwide problem regarding polymer film degradation. In the news accounts, the thin polymer film used in hundreds of clear plastic consumer products — including plastic bags (such as sandwich bags, shopping bags, dry-clean clothes covers, etc.), plastic wrap (covering such things as new CDs cases, new audio and video tapes, etc.), and even the transparent windows on mailing envelopes — have recently been exhibiting unusual properties. Consumers have reported that shopping bags break open for no apparent reason before they get to their cars! Some consumers have reported that food held in some sandwich bags has developed a foreign taste.

The U.S. Plastics Council has fielded so many complaints that Congress has been forced to act. The federal government has contacted I.O.N.S. about this problem and they want us to run some laboratory tests on a random sampling of thin plastic materials, such as those mentioned above. Since the request is for random samples, **I am asking each IONS employee to bring samples of thin plastic materials from home for this project.**

Dr. Ira Beam, a research chemist at The Poly-All Chemical Company, a major polymer film manufacturing company, is consulting with us on this. His memo and suggested procedure are attached. He is suggesting an analysis using infrared spectroscopy in order to observe possible foreign substances on or in the film.

Please prepare a memo for Senator Hanna Innevrythin, head of the Senate Committee on Consumer Affairs.

Claire

POLY-ALL

Polymer Films for Yesterday, Today, and Tomorrow

Claire:

Congratulations to you for landing this high-profile contract.

I suggest the use of infrared spectroscopy for the laboratory tests. Samples of the film can be mounted in the path of the infrared light beam in an infrared spectrometer and the resulting infrared transmission spectra recorded. If your staff is not familiar with infrared spectroscopy or the interpretation of infrared transmission spectra, you might allow them some time to read some basic reference material on this technique. I can provide that for you. The transmission spectrum recorded by the spectrometer is like a "fingerprint" of the material in the path of the light. It is a pattern that is observed each time that material is tested.

Let me further suggest testing the random samples along with polymer films that are known to be clean. Most polymer films used in shopping bags, envelope windows, CD wraps, etc., are either polyethylene or polystyrene. If you were to obtain the infrared fingerprints of polyethylene and polystyrene and compare these with the random samples, you should be able to, first, see what the polymer is, and, second, clearly see any irregularities or contaminants because they would appear as minor or major discrepancies in the patterns recorded. If the material is something other than polyethylene or polystyrene, you may be able to determine from the spectrum what it is, but you would not be able to identify foreign material without a fingerprint with which to compare it.

One other point is that these materials sometimes have an additive present to enhance the quality, and so the presence of a foreign material may not necessarily be bad. You may want to consider a second experiment in which the additive is fully identified. You may contact me for further advice on that.

So my brief procedure, which accompanies this memo, is simply to obtain infrared transmission spectra of polyethylene, polystyrene, and all the random samples. You, thus, can compare all the random samples with materials that are known to be clean and determine if there is a problem with the surface or composition of the polymers.

Ira Beam

I.O.N.S. Safety Report
prepared by Ben Whell, I.O.N.S. Safety Coordinator

Experiment 10

Equipment and Technique

- Use care with cardboard cutting tool to avoid cuts.
- Do not look directly at infrared laser source.
- Do not use a spectrophotometer that has a frayed cord.

Chemicals

- Polymer films present no safety hazard.

Workplace Cleanup

- Polymer films and waste cardboard may be recycled.
- Cardboard frames may be stored and used again.

Hazards Classifications

- Potential injuries are minor and treatable on site.

Laboratory Safety Quiz

1. What can be done to avoid cuts from the cardboard cutting tool?

POLY-ALL

Polymer Films for Yesterday, Today, and Tomorrow

Suggested SOP for the identification of
polymer films by infrared spectrometry

1. Fashion a number of mounting brackets from cardboard. The brackets should be of such size that they will fit into the slot in the infrared spectrometer's sample compartment. Cut a rectangular hole that is about 1 inch by 0.5 inch. in the center of each bracket. This is the "window" through which the infrared light beam will pass and where the polymer film will be located.

2. Attach a small piece of the film to the bracket so that it covers the window. Make sure the film is covering the entire window. You can attach the piece of film onto the bracket with tape. Be sure the film is stretched snugly across the window and there is no tape in the window.

3. Place the bracket containing the film in the instrument and obtain a transmission spectrum of the sample. Repeat for all samples to be tested.

4. Determine the identity of the samples (probably either polyethylene or polystyrene) by comparing your unknown spectra to the known spectra that have been collected. Look for patterns in the unknown spectra that don't match up with the patterns in the known spectra. This would indicate a foreign material or a material other than polyethylene or polystyrene. If the spectra don't match up with those of polyethylene or polystyrene, attempt to identify the film or the nature of the foreign material based on your knowledge of infrared spectrum interpretation protocols.

References

Kenkel J., Kelter P., and Hage D., *Chemistry: An Industry-Based Introduction*, CRC Press/Lewis Publishers, Boca Raton, FL, 2000, Chap. 6.

Hajian H. and Pescok R., *Modern Chemical Technology*, Vol. II, Prentice-Hall, Upper Saddle River, NJ, 1990, Chap. 3.

Experiment 11
MSDSs vs. Labels —
What's Missing?

I.O.N.S.
Innovative Options and New Solutions

Interoffice Memo From Claire Hemistry, CEO

Dear I.O.N.S. Staff:

I.O.N.S. has received the attached letter from Dr. Bea Saipher of the Workplace Safety and Health Administration (WSHA) in Washington, D.C. As stated in the letter, WSHA wants to have an independent evaluation of how well chemical companies and consumer products manufacturers are addressing safety issues on the labels of their products. I.O.N.S. is one independent laboratory assisting with this evaluation. WSHA has asked that all evaluators, I.O.N.S. included, utilize material safety data sheets (MSDSs) in the evaluation process.

Public safety is the primary reason for the publication of material safety data sheets. MSDSs are comprehensive descriptions which include chemical properties and hazards of which the public must be aware for the proper use, handling, storage, and disposal of these materials. The government requires that all companies that manufacture and/or distribute chemicals for use in the U.S. prepare MSDSs for those chemicals and distribute them to their customers. The government also requires that employers provide their employees with a MSDS for every chemical that is present in their workplace.

MSDSs are obviously very important documents. They provide details relating to safety well beyond that found on labels. Thus, WSHA's request that MSDSs be used in the evaluation of labels is quite appropriate. One can analyze the MSDS for particularly important safety-related information and then check the label to see if this information is adequately presented.

I have asked I.O.N.S.'s own safety coordinator, Ben Whell, to record his thoughts on this project and to tell us how to proceed. I believe he has found a useful reference that will help us understand MSDSs and labels more fully. His remarks also accompany this memo. Please prepare a memo reporting your findings for the chemicals assigned for Dr. Bea Saipher at WSHA.

Sincerely,

Claire

C. Hemistry

Workplace Safety & Health Administration

Memo to: Participating Labels Evaluation Laboratories

From: Dr. Bea Saipher

Dear Laboratory Personnel:

Thank you for participating in this year's Labels Evaluation Program. The purpose of this program is to determine if the labels that are currently being used on various manufactured chemicals and consumer products adequately address safety concerns that the public may have about them. We are confident that the Materials Safety Data Sheets that accompany these products are sufficiently complete. They were evaluated last year during the MSDS Evaluation Program sponsored by WSHA. Therefore, we recommend that the MSDSs be used as a benchmark for determining the quality of the labels. While we do require that you use the MSDSs in the evaluation process, the exact process you may use to evaluate the labels is up to you. We simply want to know if there are any crucial statements relating to safety that are missing from the labels and, if so, what statements.

Your laboratory, the I.O.N.S. Laboratory, is assigned the following industrial chemicals and consumer products:

1. Hydrogen peroxide, sold as 3% solutions both by chemical manufacturers and by drug stores as a topical anti-infective.

2. Sodium hypochlorite solution, sold as a reagent-grade solution by chemical manufacturers and as household bleach at grocery and discount stores.

3. Isopropyl alcohol, sold as a 70% solution by chemical manufacturers and as rubbing alcohol in drug and discount stores.

4. Toluene, sold as a pure reagent-grade chemical by chemical manufacturers and as a solvent by hardware stores.

I have enclosed the eight labels that are to be evaluated. You may use any current MSDS you may find for these chemicals. Again, thank you for your willingness to assist with this program. We at WSHA look forward to seeing your evaluation.

Sincerely,

Bea Saipher

Bea Saipher
WSHA

I.O.N.S.
Innovative Options and New Solutions

Memo from Ben Whell, Safety Coordinator

January 21, 2000

To I.O.N.S. staff working on the WSHA program:

I have provided your supervisors with copies of a document on MSDSs that should be very helpful as you proceed in this work. It is a section entitled "Exploring MSDSs" found in the book *Building Student Safety Habits for the Workplace,* published by Terrific Science Press at the Center for Chemical Education at Miami University, Middletown, OH. This section includes a set of instructions for doing exactly what is required for this project (Exercise 3B). I also have made photocopies of the labels in question (found in Section 2B in the above referenced book) and the MSDSs (found in Section 3B in the above referenced book). Also, your supervisors will decide who should evaluate which labels. I have recommended that each participant be assigned two of the four chemicals (four labels) at random, meaning that each participant will each have four labels to evaluate, a consumer product label and an industrial chemical product label for one chemical and a consumer product label and industrial chemical product label for a second chemical. Please use the method suggested for Exercise 3B in the book referenced above and keep a good record of your work in your notebook so that you can prepare a quality report memo for WSHA.

If you have any questions, please call.

Sincerely

Ben Whell

Ben Whell

Acknowledgment

Permission to link this experiment to the book *Building Student Safety Habits for the Workplace,* developed by the Partnership for the Advancement of Chemical Technology and published by Terrific Science Press, Miami University, Middletown, OH, is gratefully acknowledged.

Experiment 12
Mistakes at a
Steel Plant

I.O.N.S.

Innovative Options and New Solutions

Interoffice Memo From Claire Hemistry, CEO

Dear I.O.N.S. Staff:

I have just received a call from Billy Girder at the Acme Galvanized Steel Company (AGSCO). It seems that one of their customers, the Titanic Undersea Artifact Retrieval Corporation is returning most of a $1 billion galvanized steel purchase complaining that it was not up to quality standards. According to Titanic, one of their undersea artifact retrieval structures began leaking only a few months after construction was completed. They are saying that the thickness of the zinc coating on the steel was well below steel industry standards and corroded away quickly in the corrosive seawater environment.

AGSCO has determined that they stand to lose over a billion dollars if the claim is true. They have come to I.O.N.S. and have asked us to take a look at galvanized steel samples from both the galvanized sheet metal stock purchased by Titanic and also a batch of galvanized roofing nails manufactured at about the same time. They want us to determine the amount of zinc coating on the steel. They would like us to report the results both in percent zinc and also in grams of zinc per cm^2 of steel. They want us to measure both the sheet metal and the nails, since if there is a problem with the sheet metal, then there also may be a problem with the nails, since they were manufactured at about the same time.

I found an old SOP in our files from when we did a similar project for a client some years ago. I have run it by our consultants from the previous project, Dr. Ben Thair of the U.S. Department of Structural Materials (USDSM) and also Professor Hilda Van Failberg of the State University of Metals. They have both agreed to consult again on this project and have said that our SOP is appropriate. Their memoranda are attached and provide excellent guidance on this problem.

When you have finished your work, please send your report to Billy at AGSCO.

Claire

C. Hemistry

USDSM

A Note From Ben Thair

Dear I.O.N.S Staff:

The objective of this project is to determine the amount of zinc metal coating on samples of galvanized steel. Galvanized steel is steel that has had a layer of zinc (Zn) metal applied to it at the steel mill. The purpose of the zinc is to protect the steel from rusting. That is what makes the use of galvanized steel for such items as nails, garbage cans, highway guard rails, plumbing pipes, and underwater structures useful. Such materials are slow to rust and, thus, last longer.

The SOP that Dr. Hemistry has given you is for determining the percent of zinc by weight. This determination is made by weighing a sample of the steel, removing the zinc by reacting it with hydrochloric acid, and then reweighing the steel to see what the weight loss is. The amount of weight loss is the amount of zinc that was on the steel. In addition, the amount of zinc per unit area of the steel also can be determined by measuring the dimensions of a rectangularly shaped sheet metal sample and dividing the amount of zinc determined from the weight loss by the area calculated from these dimensions.

The experiment obviously requires careful weight, length, and width measurements using common laboratory measuring devices as well as subtraction and multiplication/division calculations in which significant figure rules need to be applied.

We have obtained rectangular sheet metal samples of the steel from the Titanic Corporation and the SOP is written for these. However, you will notice that the SOP does not mention nails. You will have to use your ingenuity to figure out how to make the same determinations with them.

One more thing. The industry standard for the type of galvanized steel supplied by AGSCO for this work is 6.5% zinc by weight. Galvanized steel also must meet a certain minimum standard in terms of mass of zinc per unit area. Again, for your samples, this standard is 0.00900 g/cm^2. Good Luck!

Ben Thair

State University of Metals

Providing the Keys to Success in Life

This project utilizes some very basic concepts of laboratory measurement, including the use of various measuring devices, the metric system, and significant figures. The devices used for measurement include a ruler for measuring length in centimeters and a balance for measuring mass in grams. Your supervisor will guide you in the correct use of the devices, including how to determine the number of significant figures possible from each. Remember that, unless the measuring device is digital, the reading you obtain each time a measurement is made should include all digits you know with certainty, plus one that is estimated.

Hilda Van Failberg

Professor of Steel

I.O.N.S. Safety Report

prepared by Ben Whell, I.O.N.S. Safety Coordinator

Experiment 12

Equipment and Technique

- Use care when handling metal strips with sharp edges. Wear gloves.
- Be sure to work with the HCl in a fume hood.
- Wear latex gloves when working with the HCl to avoid contact with skin.
- Use tongs when inserting the metal in, or removing it from, the acid and avoid splashing.
- Rinse the metal strips thoroughly with water before handling without gloves.

Chemicals

- The 6 M HCl gives off stifling fumes. Be sure to keep the open acid containers in the fume hood.
- The 6 M HCl is a fairly concentrated HCl solution and HCl is a strong acid. Use caution.

Workplace Cleanup

- The used 6 M HCl can be used again and again for possible future work with galvanized steel. Store in a tightly capped bottled that is clearly labeled as 6 M HCl for galvanized steel work.
- The 6 M HCl will eventually become quite discolored with dissolved zinc and iron. At that point, neutralize it with base and dispose of it according to your supervisor's instructions.
- Steel samples may be discarded or recycled.
- Rinse and dry all acid containers, tongs, etc., before storing. Discard used latex gloves.

Hazards Classifications

- Potential injuries are minor and treatable on site.

Laboratory Safety Quiz

1. What should you do in the event the 6 M HCl were spilled in the hood. What about outside the hood?
2. What should you do in the event of a cut due to the sharp edge of the metal strip?

I.O.N.S.
Innovative Options and New Solutions

SOP R992(d) — Determination of Zinc on Galvanized Sheet Metal

1. Obtain a rectangularly shaped sample of galvanized steel. Caution: the sample may have sharp edges. If necessary, clean it with warm soapy water and a brush and dry thoroughly with a paper towel. Obtain its mass in grams using the balance designated. Also obtain the length and width measurements of the steel sample (in centimeters, cm). Record all measurements in your notebook.

2. Next, remove the zinc coating by reaction with a solution of hydrochloric acid (HCl) that is labeled 6 M HCl. In a fume hood, fill a container of appropriate size with the acid to a level that will cover the entire sample of steel (Caution: Avoid Skin Contact). Your supervisor may suggest an appropriate container. Using tweezers or tongs to handle the steel sample, carefully immerse the sample in the acid. **Avoid splashing of the acid.** The reaction that takes place is

$$Zn + 2HCl \rightarrow ZnCl_2 + H_2$$

and will be evidenced by a vigorous bubbling (H_2) on the surface of the steel. The iron under the zinc also will react with the acid, but less vigorously. When the bubbling become less vigorous (should occur suddenly), the zinc has been completely dissolved and eliminated from the surface by the acid.

3. Remove the steel sample from the acid with tweezers or tongs and rinse with tap water in the fume hood sink. Continuing to handle the sample with tweezers or tongs, carry it back to your bench and rinse thoroughly with distilled water. Dry with a paper towel. When dry, weigh again on the balance. Record the mass in your notebook.

Calculations

Note: Be sure to adhere to all significant figure rules in the calculations.

1. Subtract the mass of the stripped sample from the original mass and record in your notebook. This is the mass of the zinc that was on the sample.

2. Calculate the percent of the original sample that is zinc. To do this, divide the mass of the zinc by the mass of the original sample and multiply by 100. Record the result in your notebook.

3. Calculate the surface area of the original sample in cm^2. The area of a rectangle is the length times the width. Remember that the zinc coating was on both sides of the sample, so the total area is length $\times$ width $\times$ 2. Record the result in your notebook.

4. Calculate the mass of zinc per cm^2. To do this, divide the mass of the zinc in grams by the surface area of the steel sample in cm^2. Record the result in your notebook.

5. Also calculate the mass of zinc in milligrams per m^2. To do this, convert the length and width of the steel sample to meters. Convert the mass of the zinc to milligrams. Calculate the area in m^2 and divide the mass in milligrams by the area in m^2. Record all results in your notebook.

Acknowledgment

This experiment was adapted with permission from the *Journal of Chemical Education,* 67(1), 62–63, 1990, Division of Chemical Education, Inc.

Reference

Kenkel J., Kelter P., and Hage D., *Chemistry: An Industry-Based Introduction*, CRC Press/Lewis Publishers, Boca Raton, FL, 2000, Chap. 7.

Experiment 13
The Case of the Cracked
Engine Blocks

I.O.N.S.
Innovative Options and New Solutions

Interoffice Memo from Claire Hemistry, CEO

Dear I.O.N.S. Staff:

I'm sure you will recall the news story last month concerning the great expense incurred by the City of Frozen Lakes, MN. The engine blocks of 12 of their snow removal trucks were damaged when the overnight temperature reached –30°F. I am very pleased to tell you that I.O.N.S. has won the contract to determine why the antifreeze failed.

The city's mechanics claim that a standard antifreeze formulation was used and this formulation was mixed with water in a proportion which their density instrument indicated was good to –40°F. We have obtained samples of both the formulation used and the actual antifreeze mixture found in one of their trucks. Our job will be to see if the mixture found in the truck really does have density and a composition indicating that it would, in fact, protect the truck engines.

Our consultants on this project are Dr. Ethel E. Glykle of The Great American Antifreeze Formulation Company and Professor Henry Freeze of The University of the Northern Yukon. Memos from these two outstanding scientists are attached.

Please send your report on this one to the Mayor of the City of Frozen Lakes.

Claire

C. Hemistry

The Great American Antifreeze Formulations Company

Dr. Ethel E. Glykle, Scientist

Dear I.O.N.S. Staff:

Commercial automobile radiator antifreeze formulations contain high concentration levels of two liquid organic chemicals, ethylene glycol and diethylene glycol. Glycols are organic compounds that have two hydroxyl (–OH) groups in the structure of their molecules. The structure of ethylene glycol is

$$\begin{array}{cc} \text{OH} & \text{OH} \\ | & | \\ \text{CH}_2 & \!\!\!\!\!-\text{CH}_2 \end{array} \qquad \text{ethylene glycol}$$

and the structure of diethylene glycol is

$$\begin{array}{cc} \text{OH} & \text{OH} \\ | & | \\ \text{CH}_2\text{CH}_2 & \!\!\!\!-\text{O}-\text{CH}_2\text{CH}_2 \end{array} \qquad \text{diethylene glycol}$$

Mixing an antifreeze formulation with water up to about 60% antifreeze has the effect of significantly lowering the freezing point of the water. Glycol/water mixtures, therefore, are used as the automobile radiator fluid in climates where temperatures can be expected to dip below the freezing point of water. The automobile cooling system, thus, is protected from the disabling effects of the expansion of water when it freezes, since the water/glycol mixture will not freeze. At extremely low temperatures, if the glycol concentration is high enough (50 to 60% in cold climates), the radiator fluid will not freeze even at temperatures approaching –40°F.

A method that has been used by automobile mechanics for many years to determine whether the radiator fluid in a given automobile has a high enough concentration of the antifreeze to protect the engine sufficiently is to measure the density of the fluid. The densities of glycol/water mixtures vary from 1.00 to about 1.12, depending on the concentration of the glycols. Thus, a mechanic can determine, by measuring the density, whether the mixture will protect to below the minimum temperature expected for that climate. The mechanic normally uses a device called a "hydrometer" for measuring the density.

I recommend preparing various mixtures of antifreeze and water to compare with the actual mixture sampled from the truck. Remember, the mixture found in the truck must have a concentration between 50 and 60% antifreeze in order to protect it at the temperature experienced on that night. I have written a laboratory procedure which I have attached. This procedure I believe will be useful for your efforts.

Ethel

The University of Northern Yukon

A note for you from Professor Henry Freeze

Dear I.O.N.S. Staff:

You certainly have an interesting problem to work on. I believe Dr. Glykle's procedure is a good one. Several mixtures of water and commercial antifreeze, each at a different concentration level, will be prepared and their densities measured. A graph will be prepared so as to observe graphically how the density changes as the antifreeze concentration level changes. The unknown's density will then be measured and its concentration estimated from the graph.

Actually, graphing a series of known quantities vs. corresponding concentration levels is a common technique by which to determine an unknown concentration, especially if the relationship between the measured quantity, density in this case, is linear. I would expect a straight line graph from which the unknown's concentration can easily and accurately be determined. Let me know if I can be of further assistance.

Henry F.

I.O.N.S. Safety Report
prepared by Ben Whell, I.O.N.S. Safety Coordinator

Experiment 13

Equipment and Technique

- Wear gloves to avoid contact with antifreeze and solutions.
- Wash hands after handling.

Chemicals

- Antifreeze can be harmful or fatal if swallowed.
- Antifreeze can be harmful if absorbed through the skin.
- Antifreeze can cause irritation to skin, eyes, or respiratory tract.
- Keep antifreeze away from acids and oxidizers.
- Store antifreeze in tightly capped containers.
- Keep antifreeze away from open flames.

Workplace Cleanup

- Pour antifreeze solution into designated waste container, ultimately to be disposed of via an automotive shop.
- Rinse glassware with water before storage.
- Discard used latex gloves.

Hazards Classifications

- Potential injuries include ingestion or absorption of the antifreeze.

Laboratory Safety Quiz

1. What should you do if the antifreeze comes in contact with your skin?
2. What should you do if a fellow student lights a Bunsen burner next to your work station?

Density of an Unknown Antifreeze Mixture

A laboratory procedure from the desk of Dr. Ethel E. Glykle

Caution: Wear latex gloves.

1. Obtain a clean, dry 10-mL graduated cylinder, three dry 100-mL beakers, and several clean, dry droppers. Place roughly 30 mL of distilled water in one of the beakers and roughly 30 mL of pure antifreeze in another. Label the beakers with a label indicating their contents. Label the third beaker as the "mixture" beaker. Carefully weigh the graduated cylinder on an analytical balance, being sure to obtain the correct number of significant figures. Record the weight in your notebook.

2. With one of the droppers, add antifreeze to the graduated cylinder so that the bottom of the meniscus rests on the 2.00 mL line and then transfer it to the "mixture" beaker. Allow plenty of time for this viscous liquid to run out of the cylinder and into the beaker. Rinse the graduated cylinder twice with distilled water. Discard the rinsings. Next, add water to the graduated cylinder and, with one of the droppers, add water so that the bottom of the meniscus sits on the 8.00 mL line. Also transfer this to the "mixture" beaker. Swirl the "mixture" beaker so as to *thoroughly* mix the water and antifreeze. This is a 20% (by volume) antifreeze solution. Dry the graduated cylinder with a paper towel as well as you can and then pour the mixture into the cylinder to some level between 7 and 10 mL. Carefully read the volume (to the correct number of significant figures) and record in your notebook. Now, weigh the graduated cylinder with the mixture in it on the analytical balance and also record its weight in your notebook. Discard (in the waste vessel designated) the solution in the "mixture" beaker and in the graduated cylinder. Rinse the "mixture" beaker with water and wipe it dry with a paper towel. Also rinse the graduated cylinder with distilled water and dry as best you can with a paper towel.

3. Repeat Step 2, preparing six more solutions at the 30, 40, 50, 60, 70, and 80% antifreeze concentration levels in sequence and recording the mass and volume data for each in your notebook as you did for the 20% solution. These solutions will require 3.00, 4.00, 5.00, 6.00, 7.00, and 8.00 mL of antifreeze to be mixed with 7.00, 6.00, 5.00, 4.00, 3.00, and 2.00 mL of water respectively. When all the data have been recorded, calculate the density of each solution and record in your notebook.

4. From your supervisor, obtain a sample of the antifreeze/water mixture used in the trucks and proceed to measure its density as you did the others. Record the data and results in your notebook.

5. Using graph paper, or a computer, create a graph of density on the y-axis and percent antifreeze concentration on the x-axis for your solutions from Steps 2 and 3. The plotted points should roughly show a linear behavior. With a ruler, or with the computer, draw the best straight line you can through the points. **Do not** simply "connect the dots." From the graph, determine what antifreeze concentration corresponds to the density of the unknown measured in Step 4. Record this result in your notebook.

Acknowledgment

This experiment was adapted with permission from the *Journal of Chemical Education*, 67(12), 1068–1069, 1990, Division of Chemical Education, Inc.

Reference

Kenkel J., Kelter P., and Hage D., *Chemistry: An Industry-Based Introduction*, CRC Press/Lewis Publishers, Boca Raton, FL, 2000, Chap. 7.

Experiment 14
Quality Control in
a Fluids Plant

I.O.N.S.

Innovative Options and New Solutions

Interoffice Memo From Claire Hemistry, CEO

Dear I.O.N.S. Staff:

I'm sure you've heard the recent national news reports concerning the failure of windshield washer fluid to effectively clean car windshields in six states. The problem has been traced to windshield washer fluid manufactured last month by the DO-IT-ALL FLUIDS company in Butterfly City, AZ. It seems the Federal Fluids Board (FFB) believes that the DO-IT-ALL company is deliberately diluting their windshield washer fluid with water so as to increase their profits. The company, needless to say, is very upset over the allegations and wants to clear their name in this matter. They have called upon I.O.N.S. for help.

We have signed a contract to analyze samples of DO-IT-ALL windshield washer fluid manufactured over a 10-day period last month at its plant in Butterfly City. The 10 samples were obtained under the supervision of a representative of the FFB (to avoid possible allegations that the samples are bogus) and they arrived in our labs today.

DO-IT-ALL has a quality control laboratory that monitors the specific gravity of their product daily in order to discover problems with their production. They have provided us with their SOP for this work. Your supervisor has acquired their control charting information, which was certified as genuine by the FFB. We have enlisted the help of Fred Buggs, Ph.D., as our industrial consultant, and Professor Al Cohall of DeLute University as our university consultant. Their assessment of the situation and recommendations are attached. Please pay special attention to Dr. Buggs' comments, because they explain what control charts are. Good Luck!

Sincerely,

Claire

C. Hemistry

DeLute University

College of Industrial Science

from the desk of Al Cohall, Professor of Applied Chemistry

January 20, 1998

Dear I.O.N.S. Staff:

I have been reading with a great deal of interest the news reports regarding the windshield-cleaning problems encountered across the country recently. Dr. Hemistry has described the role of I.O.N.S. in this matter and it sounds like you have a fascinating problem to solve.

I understand you have acquired certified samples of the windshield washer fluid in your lab representing the work at DO-IT-ALL FLUIDS over a 10-day period last month. You also have the SOP and certified control charting information relating to the specific gravity of their samples. I think you have all that you need to solve the case.

I would suggest that you measure the specific gravity of each of the samples and plot the control chart for this 10-day period. This will give you a visual picture of the measurements and enable you to clearly see if the samples are within the control limits for their process. The specific gravity of methanol, the main ingredient in windshield washer fluid, is very different from that of distilled water (which the FFB says is the contaminant) and, so, if water contamination is the problem, it should show up. I would also suggest that you run a control sample from a competitor whose reputation has not been called into question.

If you have any questions at any time during the analysis period, please let me know.

Sincerely,

Al Cohall

Al Cohall, Professor of Applied Chemistry

From the desk of Fred Buggs, Ph.D.

The measurement of specific gravity to check the quality of the windshield washer fluid should work very well, since even a little bit of water contamination will be noticeable. The SOP from DO-IT-ALL suggests using a pycnometer and a balance that measures mass to at least the second decimal place (the nearest hundredth of a gram).

You will need to be very careful about temperature, since specific gravity depends on temperature. The temperature of each of the 10 samples and of the water used to calibrate the pycnometer should be the same throughout all the measurements and this temperature should match the temperature at which the data for their control charting was obtained. You also will need to be concerned about warming the fluids with your hands. The volume of methanol increases dramatically with even small increases in temperature. So **handle the pycnometers as minimally as possible with your hands**. And, wear latex gloves so that you don't get fingerprints on the pycnometer which may add to the measured mass and cause an error.

DO-IT-ALL apparently routinely measures the specific gravity of their liquid products as a quality assurance procedure. I am attaching the SOP to this memo.

A **control chart** (Figure 4.1) is a graph that plots the measurement or analysis result on the y-axis and time, usually days, on the x-axis. A horizontal line corresponding to the desired value for the measurement or analysis is drawn across the center of the chart. This value is set based on a preliminary evaluation of the procedure that takes place over a

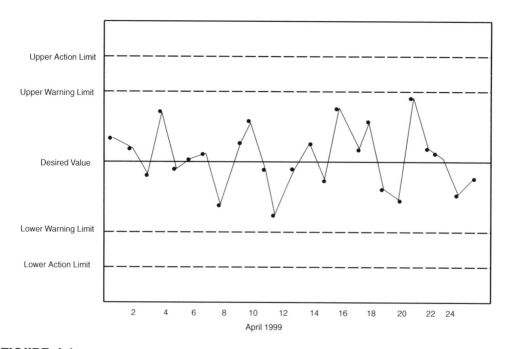

FIGURE 4.1

A control chart for a given measurement taken daily during April 1999. There are no measurements outside the warning or action limits.

period of many days. It is the statistical average produced by this preliminary work. At the same time, the preliminary evaluation produces so-called **warning limits** and **action limits** based on a statistical analysis of the data. These limits also are shown on the control chart, both above and below the desired value. Once these values have been established, the control chart is used as a day-to-day indication of whether the process is in **statistical control**, meaning that, statistically, the company's product is a quality product.

If the measurement or analysis result for a given day falls outside the warning limits, either above or below the desired value, it is not considered a cause for great concern, since, statistically, one such point is expected approximately 5% of the time, or 1 out of every 20 days. However, a point outside the action limits is cause for concern, since, statistically, such a result is expected only once out of every 333 days.

Fred

I.O.N.S. Safety Report

prepared by Ben Whell, I.O.N.S. Safety Coordinator

Experiment 14

Equipment and Technique

- Wear impervious gloves to avoid contact with chemical and solutions.
- Wash hands after handling.

Chemicals

- Methanol is poisonous.
- Methanol can be harmful if absorbed through the skin.
- Methanol can cause irritation to skin, eyes, or respiratory tract.
- Methanol is extremely flammable. Keep away from open flames, sparks, and heat.

Workplace Cleanup

- Pour windshield washer solution into designated waste container.
- Rinse glassware with water before storage.
- Discard used gloves.

Hazards Classifications

- Potential injuries include ingestion or absorption of the methanol.
- Methanol is extremely flammable.

Laboratory Safety Quiz

1. What should you do if methanol splashes on your skin?
2. What should you do if windshield washer fluid gets on your skin at home?

Standard Operating Procedure — DO-IT-ALL FLUIDS

Specific Gravity of Windshield Washer Fluid

Calibration

1. Make sure the pycnometer is dry initially, because the weight of the dry pycnometer is required in the calculations. It can be rinsed with methyl alcohol well ahead of time and allowed to dry.

2. Weigh the pycnometer (empty) with stopper to the nearest 0.01 g on a top-loading balance. This weight is the **Tare** in the calculations to follow.

3. Fill the pycnometer to the brim with distilled water that is at room temperature. This temperature also should be used for all samples. Insert the stopper slowly so that the water emerges through the hole at the top as the stopper settles in. Thoroughly wipe all overflowing water off with a Kimwipe™ being careful not to draw any water from the top of the stopper. The water level should be flush with the top of the stopper. The outside of the pycnometer should be thoroughly dry.

4. Weigh the pycnometer and record the weight on the same top-loading balance. This weight is the **Water Weight** in the calculations to follow.

Specific Gravity

1. Rinse the pycnometer twice with the sample to be tested.

2. Fill the pycnometer with the sample, using the same procedure as in Step 3 of the calibration.

3. Weigh the pycnometer with the sample on the top-loading balance. This is the **Sample Weight** in the calculations to follow.

4. To check for the reproducibility of the measurement in Step 3, empty the pycnometer, rinse and refill with the same liquid. If the weight is not the same as the weight in Step 3, you should reevaluate how you are handling the pycnometer and keep trying until you get a reproduced weight.

Calculation

$$\text{Specific Gravity} = \frac{\text{Sample Weight} - \text{Tare}}{\text{Water Weight} - \text{Tare}}$$

Results

Maintain a control chart of the specific gravity results. Use company policies to report any results that are outside the established limits of quality.

Reference

Kenkel J., Kelter P, and Hage D., *Chemistry: An Industry-Based Introduction*, CRC Press/Lewis Publishers, Boca Raton, FL, 2000, Chap. 7.

Experiment 15
Matchmaker's
Dilemma

I.O.N.S.
Innovative Options and New Solutions

Interoffice Memo From Claire Hemistry, CEO

Dear Staff:

A company called "Matchbooks Extraordinaire," which manufactures matches and match-book covers for hotels, casinos, and nightclubs, has been losing business lately and their customers have been complaining that their matches are of poor quality. These customers say that despite repeated striking of the matches, they fail to ignite most of the time. Matchbooks Extraordinaire believes the problem may stem from poor quality chemicals used in the manufacture of the matches. They especially suspect the potassium chlorate product called Matchmaker Plus that they purchase from Klein Chemicals, Inc., of Lydup City, TX.

The contract that Matchbooks Extraordinaire has with Klein calls for the Matchmaker Plus to consist of at least 50% potassium chlorate, the remaining ingredient being inert potassium chloride. They have enlisted the help of I.O.N.S. to determine the percentage of potassium chlorate in the drums of Matchmaker Plus currently at their manufacturing site. They have sent us a sample of this material for analysis.

As you may or may not know, potassium chlorate is used in match heads because when this inorganic chemical is heated to a high temperature, oxygen is released and this oxygen helps to enhance the combustibility of the match head and, thus, the match tends to light and stay lit.

I believe that we can provide Matchbooks Extraordinaire with the information they want and have enlisted the help of Max Lightner of Solid State University and Julie Wickson of Fire and Rescue International, a world-renowned fire safety consultant company. Their memos are attached. Julie was kind enough to provide us with two SOPs, one that gives a specific procedure for the use of a Bunsen burner and one that uses a Bunsen burner to decompose potassium chlorate samples and determine the percentage of oxygen in the sample. Please address your report to Claude P. Schmellsmoak at Matchbooks Extraordinaire.

Claire

C. Hemistry

SOLID STATE UNIVERSITY

Professor Max Lightner

Dear I.O.N.S. Staff:

I appreciate being able to assist with this project. The reaction in which the oxygen is released can be described by the following equation:

$$2KClO_3 + \text{heat} \rightarrow 2KCl(s) + 3O_2 \uparrow$$

This reaction is what provides the oxygen to increase the combustibility of the matchheads and it is also the reaction you will be performing in the laboratory procedure given to you by Ms. Wickson.

Actually performing this reaction in the laboratory by heating potassium chlorate in a crucible with a Bunsen burner flame does present some difficulties of which you should be aware. The $KClO_3$ has a relatively low melting point, which means that it will melt in the crucible and probably also boil. You will need to be very careful about maintaining the integrity of the sample by not allowing it to splatter out of the crucible or evaporate appreciably. Thus, please take note of the cautions mentioned in the procedure and heat slowly in the beginning.

You should also be aware that the product of the reaction, potassium chloride, has a very high melting point. Thus, as strange as it may seem, the melted, boiling liquid in the crucible will ultimately solidify despite continued direct heating. Solidification will signal nearly complete conversion of the $KClO_3$ to KCl and will mean that the final push to completion, a 15-minute direct heating period, can begin.

Good luck, and let me know if I can be of further assistance.

Max

Fire and Rescue International

August 16, 1998

Dear I.O.N.S. Staff:

I am enclosing two SOPs for your work on the Matchbooks Extraordinaire problem, one for the use of a Bunsen burner and one for the experimental determination of the percentage of oxygen. This latter procedure is written assuming that you are working with pure potassium chlorate. Of course, the material referred to as Matchmaker Plus is not pure potassium chlorate, but rather a mixture that should be at least 50% potassium chlorate.

This enclosed procedure, however, should still prove useful because the part of the Matchmaker Plus sample that is potassium chlorate will undergo the reaction and so there will be a weight loss due to the loss of oxygen from this material. The amount of potassium chlorate present can be determined by a stoichiometry calculation in which the weight of oxygen lost is converted to the weight of potassium chlorate. This weight of potassium chlorate divided by the weight of the Matchmaker Plus sample gives the fraction of Matchmaker Plus that is potassium chlorate.

If this fraction is greater than 50%, then the Matchmaker Plus meets the advertised specifications. If it is less than 50%, it does not.

One more thing. I suggest that a team of I.O.N.S. technicians, rather than just one, perform this work, if possible. This will give you results that can be averaged. Then, if one technician's data doesn't agree with the others, his/her results can be rejected as erroneous and the averaged results will be more reliable. I suggest this because there can be considerable error due to sample spattering if the sample is heated too much too quickly.

Julie Wickson

Julie Wickson, Chemist

I.O.N.S. Safety Report

prepared by Ben Whell, I.O.N.S. Safety Coordinator

Experiment 15

Equipment and Technique

- Wear gloves to avoid contact with chemical and solutions.
- Wash hands thoroughly after handling the chemicals.
- Keep containers capped at all times.
- Use caution when lighting and using the Bunsen burner. Tie back long hair.
- Be aware that objects heated with the flame (crucible, ring, clay triangle) are hot. Allow to cool before handling.

Chemicals

- Potassium chlorate is an extremely reactive oxidizer. Keep away from oxidizable materials, including organics. Consult MSDS for list of chemicals.
- Heating potassium chlorate promotes combustion by releasing oxygen. Keep combustible materials away.
- Avoid contact of potassium chlorate with eyes, skin, and clothing.
- Potassium chlorate is harmful if inhaled or swallowed.
- Keep potassium chlorate in tightly stored container.
- Potassium chloride is a skin and eye irritant.

Workplace Cleanup

- Crucible residue (potassium chloride) may be flushed down the drain with plenty of water.
- Any remaining Matchmaker Plus should be stored like potassium chlorate — away from oxidizable or combustible materials.
- Rinse glassware crucibles and spatulas with water before storage.
- Discard used gloves.
- Dispose of waste potassium chlorate as directed by local, state, and federal regulations.
- Make sure natural gas valves are completely turned off.

Hazards Classifications

- Potential injuries include burns from flames and injuries from improperly handling the potassium chlorate.

Laboratory Safety Quiz

1. What should you do if you spill potassium chlorate on the floor?
2. What should you do if potassium chlorate comes in contact with your clothes?

Fire and Rescue International

SOP — Use of a Bunsen Burner

1. The flame of a Bunsen burner is due to the combustion of natural gas, methane, in air. A Bunsen burner has independent control mechanisms for the flow of both gas and air. Examine your Bunsen burner and identify these controls. The gas control is usually a small knob that opens and closes a valve near the base of the burner. If your burner does not have such a knob, then you will have to control the flow using the valve on the benchtop. The air control consists of a mechanism whereby air is introduced into the barrel of the burner and mixed with the gas before the gas is burned.

2. With the burner gas valve closed, open the benchtop valve. Open the air controller slightly, then open the gas valve so you can hear a slight hissing sound. Light the burner with a striker or other device. If the flame is yellow, open the air controller until the yellow disappears. Now open and close the gas valve and observe the effect on the flame. With a medium-sized flame, open and close the air controller and observe the effect on the flame. With the air controller completely closed, the flame will be a bright yellow, indicating incomplete combustion of the gas. This is not a desirable condition because unburned carbon, or soot, will be deposited on crucibles and other materials that are heated using this flame.

3. As you open the controller, the yellow color disappears. However, as you continue to open the controller, you should notice a cone-shaped center to the flame begin to form. This cone is a different shade of blue than the rest of the flame. It is the hottest part of the flame. As you continue to open the air controller, this inner blue cone gets shorter and you may experience a hissing sound that is increasing in volume. Continuing to open the air controller may result in the flame being "blown out." If this happens, turn off the gas, close the air controller (not completely) and relight.

4. Set the air and gas flows to a position that gives a flame suitable for the task at hand.

Fire and Rescue International

SOP — The Determination of Oxygen in Potassium Chlorate

1. Obtain, clean, and dry a porcelain crucible and lid. Set up a ring stand with a ring and clay triangle. The ring should be at a height on the ring stand to accommodate a Bunsen burner with flame under it. Place the crucible on the clay triangle. Place the lid on the crucible, but slightly ajar to allow moisture to escape while heating.

2. Light the burner and adjust the air and fuel so as to give a hot flame at the base of the crucible. Heat the empty crucible for 15 minutes to drive off any volatile materials, then turn off the flame and allow the crucible to cool (do not remove from the triangle). The cooling should take about 15 minutes.

3. Weigh the crucible and lid together on an analytical balance and record the weight.

4. With the crucible and lid still on the balance pan, using a spatula and enough pure $KClO_3$ to the crucible so as to increase the weight approximately 1 g. Record the new weight. Calculate the weight of the $KClO_3$ and record in your notebook.

5. Place the crucible containing the $KClO_3$ back on the clay triangle (lid ajar as before) and begin heating with the burner flame again, but CAUTIOUSLY at first so that the contents of the crucible does not spatter. The reaction releasing the oxygen will proceed with a fizzing sound and a melting/bubbling appearance. Proceed cautiously until the contents has resolidified. Then, heat directly with a hot flame for 15 minutes as you did in Step 2. After this, allow to cool again for 15 minutes.

6. Weigh the crucible, contents, and lid and record the weight on the data sheet. Calculate the weight of the oxygen driven off by the heating, then calculate the percent of oxygen from the data obtained. Record all calculations and answers in your notebook.

References

Kenkel J., Kelter P., and Hage D., *Chemistry: An Industry-Based Introduction*, CRC Press/Lewis Publishers, Boca Raton, FL, 2000, Chap. 8, Sec. 8.8–8.10.

Experiment 16
Salt From Soda
Ash — Will it Work?

I.O.N.S.
Innovative Options and New Solutions

Interoffice Memo From Claire Hemistry, CEO

Dear I.O.N.S. Staff:

As you may know, sodium chloride (table salt) occurs naturally in underground salt deposits. This salt, sometimes called "rock salt," has been mined and used for many years as a source of sodium for making other vitally important sodium compounds, such as sodium hydroxide, sodium carbonate, and sodium bicarbonate. In more recent years, the soda ash deposits in Wyoming have become a direct source of sodium carbonate — soda ash is impure sodium carbonate — and, so, mining companies in Wyoming now have a significant share of the sodium carbonate market.

The Soda Ash Mining and Sales Company (SAMSCO) of Trona Ledge, WY, has proposed that the sodium carbonate made by purifying their soda ash has become a very inexpensive source of sodium and, therefore, should be used for making the other compounds rather than using the rock salt. By reacting the sodium carbonate with hydrochloric acid, sodium chloride is produced. This sodium chloride, they say, should be used to make the sodium compounds and not the sodium chloride from the rock salt.

SAMSCO believes that the yield of sodium chloride by this reaction is very high. They have come to us in order to acquire data from an independent, unbiased source to prove their point. They would like us to perform this reaction in our laboratories and determine the percent yield of the sodium chloride. They say a percent yield of greater than 90% would prove their point.

I found a short article in the *Journal of Chemical Education* that describes a useful lab procedure for this. I have written a modification of this procedure to fit this task and I have attached it to this memo. I have also attached a memo from Professor Rebecca Scrubbs of the Western Wyoming School of Mines, who provides some important background information. Unfortunately, due to the proprietary nature of the work, I was unable to locate an industrial consultant to assist.

SAMSCO has provided with us with a sample of their sodium carbonate product. Please write your report to Jimmy Prophet at SAMSCO. Thank you.

Claire

C. Hemistry

Department of Chemistry

Western Wyoming School of Mines

Rebecca T. Scrubbs, Professor

Dear I.O.N.S. Staff:

Dr. Hemistry's procedure is a straightforward one in which a given mass of solid sodium carbonate reacts with an excess of hydrochloric acid so as to quantitatively convert the sodium carbonate to sodium chloride. The reaction, indicated by Claire, is

$$Na_2CO_3(s) + 2HCl(aq) \rightarrow 2NaCl(aq) + CO_2(\uparrow) + H_2O(\ell)$$

The sodium carbonate is weighed into a preweighed porcelain crucible that was prepared in advance by washing and heating with a flame. Concentrated hydrochloric acid is added dropwise to the crucible and the above reaction begins immediately, evidenced by vigorous fizzing and bubbling, the release of the carbon dioxide. The sodium carbonate is completely reacted when there is no longer any observed reaction (fizzing) upon the addition of more HCl. At that point, an excess of HCl is added (one more drop) to be sure that the sodium carbonate is indeed the limiting reactant and is completely reacted. All that is present in the crucible at that point is the excess HCl, water, and dissolved sodium chloride. The crucible is then returned to the burner flame and all but the sodium chloride evaporated by heating. The mass of the sodium chloride product can then be determined by subtraction. The percent yield requires a stoichiometry calculation in which the mass of NaCl possible from pure sodium carbonate is calculated. This will be the theoretical yield. How much you actually do get is the actual yield.

As I see it, even if the percent yield is high, there are some obvious problems using this procedure for the bulk production of sodium chloride. I think it would be a good idea for you to ponder these and perhaps make some comments in that regard in your report to SAMSCO.

Becky

I.O.N.S. Safety Report

prepared by Ben Whell, I.O.N.S. Safety Coordinator

Experiment 16

Equipment and Technique

- Wear latex gloves to avoid contact with chemical and solutions.
- Wash hands after handling chemicals.
- Use caution when lighting and using the Bunsen burner. Tie back long hair.
- Never leave flame unattended.
- Be aware that the crucible, ring, and clay triangle may be very hot. Allow to cool before handling.
- Use fume hood when adding HCl to crucible. Use plastic dropper bottle to contain and dispense the HCl.
- Use special care when heating sodium carbonate and HCl to avoid splattering.
- Keep chemical containers closed when not in use.

Chemicals

- Sodium carbonate can cause severe eye burns.
- Sodium carbonate is harmful if swallowed or inhaled, and can cause irritation to skin and respiratory tract.
- Hydrochloric acid is a strong acid with stifling fumes. Avoid getting it on the skin. Use only in a good fume hood. Rinse thoroughly if HCl should contact the skin.

Workplace Cleanup

- Wash hands thoroughly when finished.
- Discard used gloves.
- Contents of crucible (NaCl) may be flushed down the drain.
- Make certain that the natural gas valves are completely turned off.

Hazards Classifications

- Potential acid burns if HCl contacts skin. Possible respiratory injury if fumes are not confined to fume hood.
- Potential heat burns with use of Bunsen burner.

Laboratory Safety Quiz

1. What should you do if HCl comes in contact with your skin?
2. What should you do if the sodium carbonate/HCl mixture splatters while heating?
3. What should you do if the space in the fume hood is crowded when you want to add your HCl to the crucible?

Procedure for the Determination of the Percent Yield of Sodium Chloride Formed by Reaction of Sodium Carbonate with Hydrochloric Acid

By Claire Hemistry, I.O.N.S. Corporation

1. Clean a porcelain crucible and lid with soapy water and a brush. Rinse thoroughly and dry with a paper towel. Set up a ring stand and ring with a clay triangle for heating the crucible and lid with a Bunsen burner flame as in Experiment 15 and proceed to prepare the crucible and lid as directed in Experiment 15. Allow to cool for 15 minutes and weigh.

2. Add the sodium carbonate to the crucible such that you have a sample weighing between 1 and 1.5 grams. In a fume hood, cautiously add concentrated hydrochloric acid (HCl) dropwise to the crucible and observe the reaction, a fizzing action. The reaction is

$$Na_2CO_3(s) + 2HCl(aq) \rightarrow 2NaCl(aq) + CO_2(\uparrow) + H_2O(\ell)$$

   Be careful not to add too much HCl at a time because you want to minimize loss of sample by spattering. As you continue to add the HCl dropwise, you will observe a lessening of the fizzing and eventually, even though you've added a fresh drop, there will be no observed reaction. At that point, add one more drop in excess.

3. Now take the crucible back to your station, place on your clay triangle, and begin to cautiously heat the mixture so as to evaporate the water and excess HCl. Be careful not to heat too rapidly because you do not want the water to splatter, and possibly boil over. Once the water has evaporated and only solid sodium chloride remains, heat directly with a hot flame for 15 minutes.

4. Allow to cool for 15 minutes and weigh. Subtract the weight of the empty crucible to obtain the weight of the sodium chloride.

5. Calculate the theoretical yield and then the percent yield.

Acknowledgment

This experiment was adapted with permission from the *Journal of Chemical Education,* 65(8), 731, 1988, Division of Chemical Education, Inc.

References

Kenkel J., Kelter P., and Hage D., *Chemistry: An Industry-Based Introduction*, CRC Press/Lewis Publishers, Boca Raton, FL, 2000, Chap. 8 (discussion of stiochiometric calculations, including percent yield).

Experiment 17
A Proposed New
Analytical Method

I.O.N.S.
Innovative Options and New Solutions

Interoffice Memo From Claire Hemistry, CEO

I.O.N.S. has been assigned to participate in a project sponsored by the Water Group of America (WGA) to help develop a new method for the quantitative analysis for sulfate in water samples. A procedure has already been proposed for this method and the WGA is conducting the initial investigation as to its feasibility. It is a "round robin" project in which a number of different laboratories simply check out the procedure and report back. The project is at a very early stage and they are not concerned as yet about the details (such as interferences). In fact, we have not been asked to test real water samples. They want us to take water that has a known amount of sulfate in it and see how well the procedure works — if it gives the correct answer.

Our consultant is Professor Barry M. Preesip at the Department of Chemistry, Bear Creek University. He has a special interest in water analysis and is an analytical chemist. He is very familiar with similar methods for sulfate analysis.

Please write your report to Carmen McCloudy at the WGA, but please allow me to read it before you send it on.

Claire

Claire Hemistry

Bear Creek University

From the Desk of Barry M. Preesip

Dear I.O.N.S. Staff:

I have read through the proposed procedure for the determination of sulfate in environmental water samples. It involves the precipitation of the sulfate from the water using a barium chloride solution:

$$BaCl_2(aq) + SO_4^{2-}(aq) \rightarrow BaSO_4(s) + 2Cl^-(aq)$$

This is a very well-known reaction and has been used in sulfate analysis methods in the past. What is new and different here is that spectrophotometry is used to monitor the reaction as it progresses. In that way, one can tell when the reaction is complete (when all the sulfate is consumed), how much barium chloride solution has been added to that point, and then the amount of sulfate that was present calculated via stoichiometry. The graphical picture of the progress of the reaction suggested in the procedure is useful because you can visually observe the point at which the reaction is complete and then know the mL of the barium chloride solution used at that point. This is then the starting point for the stoichiometry calculation.

The spectrophotometry measurement is based on the turbidity of the solution (cloudiness due to the dispersed precipitate). The more barium sulfate that is dispersed, the less light will pass through the cuvette, and the greater the apparent absorbance. Initially, before all of the sulfate has been precipitated, you should see an increase in this turbidity as the barium chloride solution is added. However, when all the sulfate has reacted, no more barium sulfate will form and the turbidity should not change — the absorbance readings should level off. So the graph should look something like an upside-down "L".

I see two potential problem areas with the method. One, barium sulfate is known to form at a rather slow rate. So there may be a time factor. In other words, it may make a difference if you are not consistent with the timing from one measurement to the next. If you make some measurements immediately after mixing the two solutions but measure others after 30, 60, 90 seconds, etc., the results may not be acceptable. You may want to do a set of "kinetics" experiments, i.e., do the entire experiment several times, varying the time of measurement from immediate up to 60 seconds to see if it makes a difference.

Also, the barium sulfate will not remain suspended — it will settle to the bottom of the container. This will also cause the absorbance readings to change. This means you should shake the solution just prior to measuring the absorbance. It may be a good idea to do a study on the effect of settling. This would mean not shaking the solution prior to each measurement, but measuring the absorbance at 0, 10, 20, 30 second intervals while the precipitate slowly settles.

It looks like you may have much to report to the WGA when all is said and done. Let me know if you have any questions.

Barry

Barry M. Preesip, Professor

I.O.N.S. Safety Report

prepared by Ben Whell, I.O.N.S. Safety Coordinator

Experiment 17

Equipment and Technique

- Wear latex gloves to avoid contact with chemicals and solutions.
- Wash hands after handling chemicals.
- Do not use a spectrophotometer with a frayed cord.
- Keep chemical containers closed when not in use.

Chemicals

- Barium chloride is poisonous. Just 1 g is lethal if ingested. It is harmful if inhaled, irritating to the eyes and respiratory system. It can be rendered nontoxic by precipitating the barium as barium sulfate.
- Sodium sulfate is nontoxic.
- Barium sulfate is nontoxic.

Workplace Cleanup

- Wash hands thoroughly when finished.
- Rinse glassware with water before storage.
- Discard used gloves.
- Unused barium chloride solutions may be stored for later use. Dispose of barium chloride according to state, local, and federal regulations.

Hazards Classifications

- Potential injuries can occur if ingestion or inhalation of barium chloride occurs.

Laboratory Safety Quiz

1. What should you do if barium chloride comes in contact with your skin?
2. What should you do if you spill a jar of barium chloride?

WGA Water Group of America

A Proposed New Method for the Quantitative
Determination of Sulfate in Water

Procedure

1. Switch on the spectrophotometer to be used so that it warms up. Using a volumetric flask, prepare 50.0 mL of a 0.0050 M Na_2SO_4 by diluting 5.0 mL of the 0.050 M stock to 50.0 mL. Then do the same for $BaCl_2$.

2. Obtain one 10-mL graduated cylinder (preferably with ground glass stopper), one cuvette for the spectrophotometer to be used, two droppers, and a squeeze bottle filled with distilled water. Wash the cylinder and cuvette with soap (being careful not to scratch the cuvette — use Q-tips™) and rinse thoroughly with distilled water. Prepare a two-column table in your notebook, one headed "mL of $BaCl_2$ solution" and one headed "absorbance."

3. With one of the droppers, deliver the 0.0050 M Na_2SO_4 to the graduated cylinder so that the volume in the cylinder is 3.0 mL. With the other dropper, add the 0.0050 M $BaCl_2$ so that the total volume is 3.5 mL (0.5 mL of the $BaCl_2$ solution). Now add water so that the total volume is 10.0 mL. Shake well. Rinse the cuvette with a small portion of this solution and then fill the cuvette with this solution. Wipe the cuvette with a towel or Kimwipe™ and measure the absorbance at 420 nanometers (nm) with the spectrophotometer. Record the mL of $BaCl_2$ solution and absorbance in your notebook table. Wash the graduated cylinder and cuvette with soap and rinse with distilled water, again being careful not to scratch the cuvette. If it is a disposable cuvette, you may discard it and prepare a new one.

4. Repeat Step 3 using 1.0 mL of the $BaCl_2$ solution. Continue repeating Step 3 for 1.5 mL, 2.0 mL, 2.5 mL, 3.0 mL, 3.5 mL, 4.0 mL, 4.5 mL, and 5.0 mL.

5. Plot absorbance (y-axis) vs. mL $BaCl_2$ solution on the x-axis. It should be a straight line up to where the sulfate is completely reacted. Note the mL of $BaCl_2$ solution at which the line bends. This is the mL of $BaCl_2$ solution required to react with all the Na_2SO_4 contained in 3.0 mL of the 0.0050 M solution of Na_2SO_4. Using a stoichiometric calculation, calculate how many grams of Na_2SO_4 are contained in the 3.0 mL. Check your answer by also calculating the grams of Na_2SO_4 in the 3.0 mL, knowing that the molarity is 0.0050 M.

Reference

Kenkel J., Kelter P., and Hage D., *Chemistry: An Industry-Based Introduction*, CRC Press/Lewis Publishers, Boca Raton, FL, 2000, Chap. 8, Sec. 8.11.

Experiment 18
Too Much Sodium
in Soda Pop?

I.O.N.S.
Innovative Options and New Solutions

Interoffice Memo from Claire Hemistry, CEO

December 3, 1999

Dear I.O.N.S. Staff:

Human Health International (HHI) has issued an urgent warning concerning soda pop sold by the Tasti Cola Bottling Company of St. Louis, MO, saying the soda pop has recently tested extremely high in sodium content. It is of great concern to HHI because Tasti Cola's products are so popular among consumers.

The Tasti Cola people are alarmed at the warning because their quality assurance department has never found any problem with any of their products, and they perform routine sodium tests on a daily basis. They have been told by HHI that an independent testing laboratory, the We Analyze Anything (WAA) laboratory of Chilly Falls, WI, tested the products and have reported the stunning results. Tasti has come to I.O.N.S for yet another independent test to confirm their own results and refute the WAA results. We have agreed to a contract to perform the required tests. We have the original undiluted samples tested by WAA.

I have secured the services of Abby Zorb, Ph.D., as our industrial consultant and Professor Pete Pehem as our academic consultant for this work. Their comments, as well as FSPA SOP 46C suggested by Dr. Zorb, accompany this memo. Please send a memo reporting your results to Phil Thahkann at Tasti Cola. Good Luck!

Sincerely,

Claire

Claire Hemistry

from the desk of Professor Pete Pehem

November 19, 1999

Claire:

Thanks for the opportunity to assist with this project. It is a very interesting problem and one that I'm sure your chemists and chemistry technicians can solve.

The analysis technique your laboratory should choose for this is called "atomic absorption spectroscopy." It is a technique that utilizes a fairly sophisticated instrument that you have in your laboratory. It is frequently employed for quantitatively determining metal ions in solution. In fact, sodium is easily determined with this technique. It involves introducing the solutions to be tested into a flame and measuring the amount of light absorbed by the sodium atoms in the flame. The light to be absorbed originates from a special lamp called the sodium hollow cathode lamp. The atoms absorb light when their electrons acquire energy from the light and are elevated to a higher energy level. The amount of light absorbed is measured by the instrument and a number representing this absorption (the "absorbance") is displayed on a readout meter. The more sodium in your samples, the higher the absorbance. This absorbance can be compared to a series of standard solutions of sodium and the amount of sodium in the soda pop calculated.

Good luck and please let me know if I can be of further assistance.

Pete

From the Desk of Abby Zorb, Ph.D.

The most critical parts of the laboratory work you are about to undertake are the preparation of the soda pop sample and the preparation of the standards. I suggest the Federal Soda Pop Administration (FSPA) SOP 46C. In this procedure, the soda pop is prepared for analysis by first removing the carbonation and then diluting the remaining liquid. This dilution, 1.00 mL of soda popa diluted to 25.00 mL of solution, is done to lower the concentration of sodium so that it is within the range of the standards to be prepared. The standards are prepared by diluting a stock standard that is a 1000 ppm solution of sodium, first to make a 100 ppm solution, and then to make solutions that are 1.00, 3.00, 5.00, and 7.00 ppm by diluting the 100 ppm.

The absorbance values obtained from the instrument are plotted vs. the ppm of the standards. The diluted soda pop concentration is then determined from this graph. To get the amount of sodium in the undiluted sample, you will need to multiply the ppm obtained from the graph by 25, since 1 mL of the soda pop was diluted to 25 mL.

HHI has declared that the safe upper limit for human consumption of sodium in soda pop is 200 ppm. In other words, if the calculated result for your sample exceeds 200 ppm, then HHI's fears are realized and there is a problem with the soda pop.

Good luck, and if you have any questions, please call on me.

Abby

I.O.N.S. Safety Report

prepared by Ben Whell, I.O.N.S. Safety Coordinator

Experiment 18

Equipment and Technique

- Wear latex gloves to avoid contact with sodium solutions.
- Do not use an ultrasonic bath with a frayed cord.
- The atomic absorption unit should not be operated without proper supervision. Operate according to instructions and use caution with flame and in handling acetylene. Be sure exhaust hood over the instrument is on and the drain line is properly configured.
- Use special care when installing the cathode lamp to avoid breakage or electrical hazard.
- Keep solution containers closed when not in use

Chemicals

- Sodium standard is a relatively harmless solution that can cause a slight irritation to the eyes, respiratory system, and respiratory tract.
- Acetylene is a highly flammable gas. Be certain that the flame is lit when the valves are open and that there are no leaks in the line.

Workplace Cleanup

- Wash hands thoroughly when finished.
- Rinse glassware with water before storage.
- Discard used gloves.
- Make sure acetylene is turned off both at the tank and at the instrument.

Hazards Classifications

- Potential injuries can occur from burns from the atomic absorption flame.
- Backflash can occur from improper mixtures of fuel and oxidant or if drain line is not properly configured.

Laboratory Safety Quiz

1. Why isn't it sufficient to just close the acetylene valve on the instrument and not at the tank?

Federal Soda Pop Administration

SOP 46C — Analysis of Soda Pop for Sodium

1. Using a volumetric flask, prepare 25 mL of a 100 ppm stock sodium solution from the available 1000 ppm solution.

2. From the 100 ppm stock solution, prepare standard solutions that are 1.00, 3.00, 5.00, 7.00, and 9.00 ppm in 25 mL volumetric flasks. Use a pipetter to measure the required volumes of the 100 ppm solution.

3. Degas the soda pop sample by placing 10 to 20 mL in a small beaker and then placing the beaker in an ultrasonic bath for a few minutes.

4. Using a pipetter, pipet 1.00 mL of the soda pop into a 25-mL volumetric flask and dilute to the mark with distilled water. Shake.

5. Obtain absorbance values for all standards and sample with an atomic absorption instrument.

6. Plot absorbance vs. ppm and obtain the ppm concentration of sodium in the diluted sample. Then, multiply this concentration by 25 to get the ppm concentration of sodium in the original soda pop since you diluted it by a factor of 25 in Step 4.

References

Kenkel J., Kelter P., and Hage D., *Chemistry: An Industry-Based Introduction*, CRC Press/Lewis Publishers, Boca Raton, FL, 2000, Chap. 10 (discussion of dilution and dilution calculations).

Experiment 19
Conductivity, Odors, and Colors — Oh My!

I.O.N.S.
Innovative Options and New Solutions

Interoffice Memo From Claire Hemistry, CEO

Dear I.O.N.S. Staff:

The We Analyze Anything (WAA) Laboratory in Chilly Falls, WI, was so impressed with our previous soda pop analysis work (Experiment 18) that they have actually become our client. You will recall that their analysis of soda pop samples from Tasti Cola Bottling Company indicated unusually high levels of sodium. You also will recall that we proved them wrong with our own analysis.

Fearful that they may lose their reputation in the community, WAA requested a government audit of their operations to see if the problem could be internal. Indeed, the auditor concluded that the problem was likely due to a contaminated distilled water supply. It turns out that WAA had previously been using a conductivity meter to continuously monitor the quality of the water as it elutes from the distiller and had taken it offline some time earlier because it was apparently malfunctioning. Besides always indicating that the water was highly conductive despite the use of an expensive distillation and deionizing system, it also seemed to be the source of unusual colorations and odors found in the water. As soon as it was taken offline, the colorations and odors ceased.

The auditor's conclusion that the water supply is contaminated has led us to suspect that WAA's conductivity meter was actually functioning properly and that the problem is either in the distillation/deionizing equipment or in the untreated water. To help prove this theory, we decided to follow the suggestion of Professor Ely Pewtred of Foulwater University and test a series of solutions prepared using our own water and then use our own conductivity equipment to see if we could determine what may be giving rise to the colorations and odors.

Please follow Dr. Pewtred's advice. His correspondence accompanies this memo.

Claire

Professor Ely Pewtred

THE FOULWATER UNIVERSITY

Dear I.O.N.S Staff:

I'd be willing to bet WAA's problem is either in the distillation/deionizing equipment or in the untreated water. There are quite a number of solute chemicals that will produce the colorations and odors that WAA experienced with the water that had been passed through the online conductivity meter. To test our theory that one or more of these chemicals are contaminating the water supply prior to reaching the meter, I have devised a procedure in which you prepare a solution of each of these chemicals and then test all of them with a conductivity apparatus, one in which you can see and smell the substances produced by the process.

I am providing some background information on conductivity testing and electrolysis processes. This information should help you understand the theory of the testing, what processes occur, and how these processes can produce odors and colors.

Of course, the work requires that you prepare a number of solutions that contain the solute chemicals. I have included a worksheet you can use to organize the calculations required. Depending on the wishes of your supervisor, you can either use the worksheet as a guide for your laboratory notebook or only place examples of the calculations in your notebook. At any rate, you should keep good records of the grams (if you are preparing the solution using a pure, solid chemical) or milliliters (if you are diluting a more concentrated solution) you determine from the calculation so that any apparent errors can be traced. You also should record the conductivity of each solution measured with the suggested apparatus, as well as any colors and odors that you observe. In the end, you should be able to suggest possible contaminants in the water, those chemicals that gave you colors and odors.

Ely

Ely's Background Information

Some chemical substances, when dissolved in water, break apart into positive and negative ions such that free ions exist in solution and give the solution certain characteristic properties. One very important such property is the ability of the solution to conduct an electric current. Solutions that contain ions will conduct an electric current. Solutions that do not contain ions will not conduct an electric current. This property is the basis for a number of laboratory and industrial processes, including the manufacture of certain chemicals, electroplating of metals, and the quantitative analysis of dissolved chemical species.

The term "electrolyte" is used to describe a substance that, when dissolved in water, breaks apart into ions and, therefore, conducts an electric current. This term is also used to describe the solution itself.

Since the dissolving of certain substances can result in either complete or incomplete ionization, as well as no ionization at all, additional terms are used to describe these. These terms are: strong electrolyte (complete ionization), weak electrolyte (partial ionization) and nonelectrolyte (no ionization).

As a general rule, substances that are held together by ionic bonds (ionic compounds) are electrolytes while substances that are held together by covalent bonds (covalent compounds) are nonelectrolytes. The exceptions to this general rule are acids because they are covalent compounds whose solutions conduct an electrical current. Strong acids are strong electrolytes and weak acids are weak electrolytes. Strong acids (e.g., HCl, H_2SO_4, and HNO_3) completely ionize when dissolved in water and, thus, even though they are covalent compounds, they are electrolytes. Also, some ionic compounds do not dissolve in water (precipitates) and, thus, there are no ions in solution in that case.

The electrolyte properties of a solution can be measured by a rather simple apparatus consisting of two electrodes, such as graphite rods, dipped into the solution, a battery, and current measuring device (ammeter) all connected with wires or electrical "leads" (Figure 4.2). If a current is able to flow through the solution, it will register on the ammeter. If the solution cannot conduct a current, the meter will read zero.

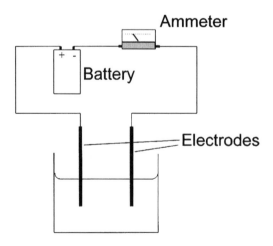

FIGURE 4.2
The apparatus for measuring the electrolyte properties of solutions.

Let us examine what happens in the solution when a current flows. First, the presence of ions is a requirement for current flow. This is because the movement of charge (current) through the solution can only result from the movement of ions. The positive ions diffuse to the negative electrode (connected to the negative pole of the battery), while the negative ions diffuse to the positive electrode (Figure 4.3). If there are no ions present, there is no current flow and the solution would be a nonelectrolyte.

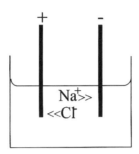

FIGURE 4.3
When the conductivity apparatus is in operation and ions are present in the solution, the positive ions diffuse toward the negative electrode and the negative ions diffuse toward the positive electrode.

Secondly, in addition to the need for ions to be diffusing through the bulk of the solution, there must be some mechanism for consuming electrons at one electrode and generating them at the other. This is because the charge movement constituting the current flow through a metal conductor (the electrode and the connection to the battery) consists of the movement of electrons. Electrons move to and from the battery (Figure 4.4).

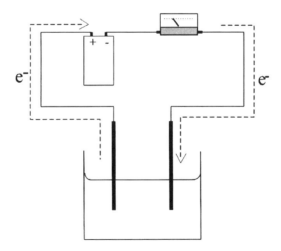

FIGURE 4.4
When a current flows through the circuit, electrons flow to and from the battery.

The questions then become: what happens to the electrons flowing down the negative electrode and where do they come from when they flow up the positive electrode? **Electrons cannot flow on their own through the solution.** The answer is that oxidation and reduction processes occur at the surfaces of the two electrodes. At the surface of the negative electrode, electrons are captured by a chemical species in contact there (e.g., the Na^+ ions), while at the positive electrode, electrons are given up to the surface of the electrode by another chemical species in contact there (e.g., the Cl^- ions). In other words, reduction of some chemical species takes place at the negative electrode while oxidation of another chemical species takes place at the positive electrode, thus completing the circuit, allowing electrons to flow freely to and from the battery and registering a current on the ammeter. Using the Na^+ and Cl^- as examples, we can write the following equations to indicate the two processes using "e^-" to symbolize an electron:

$$Na^+ + e^- \rightarrow Na$$

$$2Cl^- \rightarrow Cl_2 + 2e^-$$

Often, there is physical evidence of the oxidation/reduction processes. For example, for a solution of NaCl giving the reaction written above, vigorous gas bubbling can be observed occurring at both electrodes. The gas forming at the positive electrode is chlorine, Cl_2. The odor of chlorine can be detected after a very short time. At the negative electrode, the gas is hydrogen resulting from the reaction of sodium metal with water.

Besides observing gases evolving at the two electrodes, as with a NaCl solution, other observations are made with other dissolved chemicals. For example, if an iodide (I^-) or bromide (Br^-) salt is dissolved in the solution, the following reactions occur at the positive electrode:

$$2I^- \rightarrow I_2 + 2e^-$$

$$2Br^- \rightarrow Br_2 + 2e^-$$

In these cases, a discoloration of the solution near the electrode surface is observed after a few seconds indicating the formation of the I_2 or the Br_2. Interesting reactions also may occur at the negative electrode. If a copper salt or a silver salt is dissolved, copper metal or silver metal will be observed "depositing," or "electroplating" on the electrode surface because of the following reactions:

$$Cu^{+2} + 2e^- \rightarrow Cu$$

$$Ag^+ + e^- \rightarrow Ag$$

In summary, the flow of current through a solution occurs because (1) ions diffuse through the bulk of the solution and (2) electron transfer occurs at the surfaces of the electrodes. If any of these processes are absent, current will not flow.

I.O.N.S. Safety Report

prepared by Ben Whell, I.O.N.S. Safety Coordinator

Experiment 19

Equipment and Technique

- Wear latex gloves to avoid contact with chemicals and solutions.
- Wash hands after handling chemicals.
- When preparing solutions of acids or bases, place water in the container first.
- Keep chemical containers closed when not in use.
- Keep ethanol away from sparks, heat, and flames.

Chemicals

- Acids and bases are corrosive and can cause severe eye and skin burns.
- Copper sulfate is harmful if swallowed. It is irritable to the eyes and skin.
- Potassium iodide may cause irritation upon contact with eyes, skin, and clothing.
- Sodium bromide may cause irritation upon contact with eyes, skin, and clothing.
- Ethanol is flammable.
- The 0.50 M solution of ammonium hydroxide will have an ammonia odor. Be cautious when handling.
- Sodium hydroxide is a strong base. It can cause severe chemical burns.

Workplace Cleanup

- Wash hands thoroughly when finished.
- Rinse glassware with water before storage.
- Discard used gloves.
- Dispose of solutions according to local codes and chemical hygiene plan.

Hazards Classifications

- Potential injuries are limited and locally treatable.

Laboratory Safety Quiz

1. What would you do to clean up a spill of 2.5 M sulfuric acid?
2. Why do you add acid to water instead of water to acid?

Ely's Procedure

Part A. Preparation of Solutions

One task to be performed in this assignment is the preparation of a series of solutions. Most of these solutions are of a particular molarity. These will be prepared either by diluting solutions of higher molarity or by weighing pure solid chemicals and dissolving and diluting to volume. Other solutions are of a particular percent, either volume percent (liquid solute) or weight/volume percent (solid solute). These will be prepared either by weighing the solid solute and dissolving and diluting to volume, or by measuring the volume of the liquid solute and dissolving and diluting to volume. To determine the weights and volumes of solutions and chemicals to be used for the solutions, you will use the formulas for dilution, molarity, and percent derived from your previous experience.

Eleven solutions are needed. The worksheet can be used to calculate the amount of the required chemical or solution needed for each solution. Prepare 100 mL of each in 125-mL Erlenmeyer flasks that have a calibration line for 100 mL. Record the amount calculated on the worksheet and in your notebook.

Part B. Observation of Electrolyte Properties

Measure the conductivity of each solution and also the I.O.N.S. tapwater and distilled water using the apparatus described in the reference material. Your supervisor will show you how to assemble and use it. Be sure to **rinse the container and electrodes with distilled water after each measurement** so that there is no contamination of any solution. Also be sure to **fill the container to the same depth each time**. The more electrode surface exposed to the solution, the higher the current. Thus, filling the container to different levels may lead to false conclusions.

In addition to recording the current readings in your notebook, also record your conclusion as to whether the solution tested is a strong, weak, or nonelectrolyte, and any observations of odor, colorations, or electroplating that may occur.

Ely's Worksheet

Solution 1: 0.10 M acetic acid. You are to dilute a stock solution that is 1.0 M.

mL of 1.0 M acetic acid used _____

Solution 2: 0.10 M NaOH. You are to weigh out the appropriate number of grams of the pure, solid NaOH.

grams of NaOH to be weighed _____

Solution 3: 0.10 M HCl. You are to dilute a stock solution that is 2.0 M.

mL of 2.0 M HCl to be used _____

Solution 4: 0.10 M NaCl. You are to weigh out the appropriate number of grams of pure, solid NaCl.

grams of NaCl to weigh _____

Solution 5: *0.10 M NH₄OH. You are to dilute a stock solution that is 0.50 M.*

mL of 0.50 M NH$_4$OH to be used _____

Solution 6: *0.10 M KI. You are to weigh out the appropriate number of grams of pure,*
solid KI.

grams of KI to weigh _____

Solution 7: *A solution that is both 1.0 M CuSO₄ · 5H₂O and 0.10 M H₂SO₄. The CuSO₄ ·*
5H₂O is a pure solid and the H₂SO₄ is from a stock solution that is 2.5 M.

grams of CuSO$_4$ · 5H$_2$O _____

mL of 2.5 M H$_2$SO$_4$ _____

Solution 8: 0.010 M HCl. You are to dilute the 0.10 M HCl solution (Solution 3).

mL of 0.10 M HCl _____

Solution 9: 5.0% (volume percent) ethyl alcohol. You are to measure the appropriate volume of the liquid ethyl alcohol.

mL of ethyl alcohol _____

Solution 10: 0.10 M NaBr. You are to weigh the appropriate number of grams of pure, solid NaBr.

grams of NaBr to weigh _____

Solution 11: 5.0 % (weight/volume percent) sugar. You are to weigh the appropriate number of grams of sugar.

grams of sugar to weigh _____

Reference

Kenkel J., Kelter P., and Hage D., *Chemistry: An Industry-Based Introduction*, CRC Press/Lewis Publishers,
    Boca Raton, FL, 2000, Chap. 10.

Experiment 20
Humid Fun for National Chemistry Week

I.O.N.S.
Innovative Options and New Solutions

Interoffice Memo from Claire Hemistry, CEO

Dear I.O.N.S. Staff:

The local section of the American Chemical Society (ACS) is preparing for the next National Chemistry Week (NCW). This is a week set aside each year in November during which the ACS, with its 159,000 members, carries out a public relations blitz to improve the public image of chemistry as a discipline, as a field of study, and as a career. While we chemists know of chemistry's role in the modern world as the "central science" that is deeply concerned with the common good and economic progress, the general public remains fixated on the negative impressions — the perceived dangers from such things as pollution, holes in the ozone layer, hazardous wastes, and oil slicks, all presumably caused by chemistry. Hence, the need for the annual blitz. NCW has been held annually since the late 1980s and is usually approached with much enthusiasm by ACS members and other professionals in the field.

In order to get elementary and high school students turned on to chemistry, local ACS members like to visit the schools and stage chemistry demonstrations and activities during the week. This year, our local community college, Marie Curie Community College (MCCC), has selected an activity that is sure to please — a paper wall hanging that utilizes a chemical equilibrium to indicate the relative humidity of indoor air.

MCCC would like to guarantee the success of this activity by assuring that the wall hanging will work as well as it can. They have a procedure to follow, but they also would like to see if it can be improved. They have come to us with the problem to see if we can help optimize the quality so that it truly is a good indicator of humidity.

Of course, this work is completely voluntary on our part and you may not take an extraordinary amount of time away from our usual work. Still, we want to play our part for the success of the local endeavor. Please refer to the memo from my brother, Clyde Hemistry, who is an instructor in the chemistry program at MCCC.

Sincerely,

Claire

I.O.N.S. Safety Report
prepared by Ben Whell, I.O.N.S. Safety Coordinator

Experiment 20

Equipment and Technique

- Wear latex gloves to avoid contact with chemical and solutions.
- Wash hands thoroughly after handling chemicals.
- Use caution when handling filter paper saturated with cobalt solution.
- Keep chemical containers closed when not in use.

Chemicals

- Cobalt nitrate is a strong oxidizer, harmful if swallowed or inhaled, irritable to eyes, skin, and respiratory system. Store away from oxidizable material.
- Cobalt chloride is harmful if swallowed, can cause irritation to skin, eyes, and respiratory tract.
- Ammonium chloride causes irritation to skin, eyes, and respiratory tract and is harmful if swallowed or inhaled.
- Sodium chloride is an eye irritant.
- Hydrochloric acid is a strong acid with stifling fumes. Use plastic dropper bottle. Use in a fume hood.
- Inform MCCC faculty that students should wear gloves and wall hangings should be covered with clear plastic.

Workplace Cleanup

- Wash hands thoroughly when finished.
- Rinse glassware with water before storage.
- Discard used gloves.
- Dispose of cobalt solutions according to state, federal, and local regulations, being aware that some contain ammonium chloride and sodium chloride.
- Dispose of cobalt-saturated filter paper according to state, federal, and local regulations.

Hazards Classifications

- Potential injuries are limited and locally treatable.

Laboratory Safety Quiz

1. Why should you not eat or drink in a laboratory?
2. What should you do if an excess of cobalt nitrate gets on your clothes?
3. What would happen if cobalt nitrate came in contact with an oxidizer?

MCCC

Dear I.O.N.S. Staff:

Thank you so much for your willingness to volunteer to help with this event. National Chemistry Week has grown so much in the past few years and we want to be part of its continued growth and success.

We have found a simple procedure for creating the wall hangings. It uses the following equilibrium reaction. The $Co(H_2O)_6^{2+}$ species is pink in color and the $CoCl_4^{2-}$ species is blue.

$$Co(H_2O)_6^{2+}(aq) + 4Cl^-(aq) \rightleftharpoons CoCl_4^{2-}(aq) + 6H_2O(\ell)$$

$$Pink \rightleftharpoons Blue$$

The paper used in the wall hanging is soaked with a solution of $CoCl_2 \cdot 6H_2O$ and allowed to dry. The solution, and thus the wall hanging, has all four of the species involved in the reaction in it. The theory is that on a humid day, with Le Chatelier's Principle in effect, water from the humid air adsorbs on the paper and pushes the equilibrium to the left changing the color from blue to pink.

Addition of water (high humidity):

$$Co(H_2O)_6^{2+}(aq) + 4Cl^-(aq) \longleftarrow CoCl_4^{2-}(aq) + 6H_2O(\ell)$$

$$Pink \longleftarrow Blue$$

On a dry day, there is less H_2O and the equilibrium is pushed to the right making the paper blue.

Removal of water (low humidity):

$$Co(H_2O)_6^{2+}(aq) + 4Cl^-(aq) \longrightarrow CoCl_4^{2-}(aq) + 6H_2O(\ell)$$

$$Pink \longrightarrow Blue$$

Temperature has a small effect, too.

The recipe for the soak solution was obtained from the *Journal of Chemical Education*'s December 1991 issue. You should weigh 10 g of $CoCl_2 \cdot 6H_2O$ and dissolve it in 25 mL of water. The material for the wall hanging is Whatman #2 qualitative filter paper.

The idea we had was to do a little research to see if the color swings could be a little better defined over a broader humidity range, thus being a little more precise in indicating the humidity value. This research would involve a series of trials involving adding and taking away chloride ions from the equilibrium. Perhaps either fewer chloride ions or more chloride ions would produce better color changes when the humidity changes. Chloride ions are certainly involved in the equilibrium, and adding them or taking them away would certainly affect the color. *Chloride ions added:*

$$Co(H_2O)_6^{2+}(aq) + 4Cl^-(aq) \rightarrow CoCl_4^{2-}(aq) + 6H_2O(\ell)$$

$$Pink \rightarrow Blue$$

Chloride ions removed:

$$Co(H_2O)_6^{2+}(aq) + 4Cl^-(aq) \leftarrow CoCl_4^{2-}(aq) + 6H_2O(\ell)$$

$$Pink \leftarrow Blue$$

To add chloride ions, you can dissolve some NaCl or NH$_4$Cl crystals in the cobalt chloride solution in varying amounts (say, from 0 to 5 g) and soak a piece of filter paper (perhaps cut to a size that is about 1 in.^{2}) after each addition. To take away chloride ions, begin with cobalt nitrate, Co(NO$_3$)$_2 \cdot$ 6H$_2$O, instead of the cobalt chloride, such that there are no chloride ions. This would be one solution in which to soak a piece of filter paper. Then add NaCl or NH$_4$Cl crystals in varying amounts and soak the paper after each addition.

There are lots of possibilities, but the following could be one set in which 10 different test squares would be created.

Using cobalt chloride — 10 g per 25 mL of water:

1. Nothing added
2. 1.0 g NaCl added
3. 1.0 g NH$_4$Cl added
4. 2.0 g NaCl added
5. 2.0 g NH$_4$Cl added

Using the cobalt nitrate solution — 10 g (or other amount) per 25 mL of water:

1. Nothing added
2. 1.0 g NaCl added
3. 1.0 g NH$_4$Cl added
4. 2.0 g NaCl added
5. 2.0 g NH$_4$Cl added

The first set of five would test the effect of more chloride than the originally recommended solution and the second set would test the effect of less chloride than the originally recommended solution.

When finished preparing and soaking the filter papers, you can place them on watch glasses in a fume hood to dry, then later hang them in the lab (maybe tack them on the lab bulletin board) and monitor the colors each day for a week or more to observe the color changes as the humidity changes. Monitoring the humidity with an electronic humidity indicator alongside the filter papers would be important in order to correlate the color changes with the actual humidity swings. The filter paper that demonstrates the most noticeable color swings over the range of humidity values tested should represent the best recipe for the soak solution.

By the way, sometimes the indoor humidity doesn't vary much over a week's time. If you find this to be the case, you can create artificial humidity swings by using a 10- or 20-gal fish tank that has a 1-in. deep layer of water on the bottom. If you place the filter papers inside the fish tank (perhaps taped to the side and place the humidity meter in the fish tank, too) and cover it, you can create higher humidity values.

I might suggest some preliminary studies just to see what the colors are as the equilibrium shifts in the different directions. To do this, I would suggest preparing a solution of cobalt nitrate, perhaps 0.4 M, and then adding chloride ions either by dropwise addition of concentrated HCl, or by adding crystals of NaCl. You should see the shift from pink to blue.

May I also suggest one final test? Since there is some effect of temperature, perhaps you can take the $Co(NO_3)_2$ solution to which has been added some 12 M HCl (about a 5:3 ratio of cobalt solution and the HCl) and split it into three separate test tubes. Heat one test tube in a hot water bath and cool another in an ice bath. This should result in a visible shift in equilibrium and, therefore, pink/blue color changes. From these observations, you should be able to tell which reaction (forward or reverse) is favored by more heat.

Good Luck! I will be looking forward to your findings.

Clyde

Clyde Hemistry

Acknowledgment

This experiment was adapted with permission from the *Journal of Chemical Education,* 68(12), 1039, 1991, Division of Chemical Education, Inc.

Reference

Kenkel J., Kelter P., and Hage D., *Chemistry: An Industry-Based Introduction,* CRC Press/Lewis Publishers, Boca Raton, FL, 2000, Chap. 11.

Experiment 21
The Soil
Project

I.O.N.S.

Innovative Options and New Solutions

Interoffice Memo From Claire Hemistry, CEO

Dear I.O.N.S. Staff:

The National Soil Conservation Administration (SCA) has extended an invitation to the I.O.N.S. Corporation to participate in a round robin proficiency testing activity. This is an activity in which a laboratory demonstrates its proficiency in performing certain functions in a statistically accurate manner and, thereby, establishes credibility with its clients. Other laboratories also will be participating. Our results will be compared to that of other laboratories as well as to the results expected by the SCA. The SCA will then grade our proficiency.

We have received three soil samples from the SCA. The instructions call for us to analyze these samples for moisture, calcium carbonate, pH, and mineral content. The SCA has provided the procedures for these analyses.

So that you have full background knowledge of these procedures, I have asked Professor Dusty Fields of the Agronomy Department at Farmer University to consult. He has reviewed the Soil Conservation Administration Proficiency Testing Program procedures and has presented us with much useful information in his attached memo. Please prepare your report memo for Phyllis W. Vejeese at SCA.

Claire

Farmer University

Helping the world grow better crops since 1928

Dear I.O.N.S Staff:

Thanks for asking me to consult on these test procedures. Here are my thoughts. Please correlate my remarks with the procedure from the SCA.

Moisture. The temperature of 105°C noted in Step 1 indicates that the soil samples do not contain organic matter that volatilizes readily. Normally, a temperature of 70° or lower is used in that case so that the weight loss includes only the moisture and no organic matter. Otherwise, this is a simple, straightforward procedure and your lab should be able to be quite accurate with it.

Sample Preparation. It is important for the subsequent tests to maximize the surface area exposed to the water and solutions to be added. Otherwise, there may be some reactive material inside a particle that would not be able to react as needed simply because it is not on the surface. Hence, the need to dry, crush, and sieve. The 2-mm size is a pretty standard size for these kinds of tests.

Carbonate Spot Test. The presence of carbonates in a soil is important because carbonates affect the properties of a soil. Such properties include the cation exchange capacity and the water holding capacity. It is important for a soil scientist to determine if carbonates are present and to also quantitatively analyze for carbonates. The carbonate spot test determines if carbonates are present. The reaction of carbonates with acids yields carbon dioxide gas and, thus, an effervescence is observed upon contact. The reaction of hydrochloric acid with calcium carbonate is:

$$CaCO_3(s) + 2HCl(aq) \rightarrow CaCl_2(aq) + CO_2\uparrow + H_2O(\ell)$$

Quantitative Carbonate Test. This is a good quantitative test for calcium carbonate. There should be a weight loss because the carbon dioxide is released into the air. This weight of carbon dioxide can then be converted to the weight of calcium carbonate using a stoichiometry calculation with the above reaction. The % calcium carbonate is calculated by dividing the weight of calcium carbonate by the weight of the soil sample.

Soil pH. The pH of a soil controls the ability of a plant to grow, as well as how soluble aluminum and other heavy metals are in the soil solution and how well microorganisms can grow in the soil. Thus, it is a critical measurement for soil scientists. Soil pH is measured by creating a soil solution with either distilled water, 1 M KCl, or 0.01 M $CaCl_2$ added to a sample. I see that you are using the distilled water method here. This is the simplest method and is good enough for many soils, but it is not desirable for soils that vary or are high in salt content. For these, the others are more desirable because they mask the salt content. They do, however, result in pH values different from the distilled water test in these cases because hydrogen ions are displaced from the soil's cation exchange sites.

Mineral Content. The mineral content consists of the material remaining after organic matter is removed. The organic matter is driven off when the soil is "ignited" or heated to a very high temperature, such as 400°C or higher. The loss of organic matter is often called "loss on ignition" and defines organic soils.

This should be a fun and exciting project for you. I am available in case you need to contact me.

Sincerely,

Dusty Fields

Professor Dusty Fields

I.O.N.S. Safety Report
prepared by Ben Whell, I.O.N.S. Safety Coordinator

Experiment 21

Equipment and Technique

- Wear latex gloves to avoid contact with chemicals and solutions.
- Wash hands after handling.
- Use caution when lighting and using Bunsen burner. Long hair should be tied back.
- Do not leave Bunsen flames unattended.
- Keep chemical containers closed when not in use.
- Use tongs or hand insulators to handle hot objects.
- Do not use electrical appliances with frayed cords.

Chemicals

- Hydrochloric acid is a strong acid. The 6.0 M concentration is especially dangerous. If skin contact occurs, rinse immediately. Also be aware of stifling fumes. Fill vials in a fume hood and keep vials capped when not in use. If spills occur, neutralize with a solution of a weak base.

Workplace Cleanup

- Wash hands thoroughly when finished.
- Rinse glassware with water before storage.
- Discard used gloves.

Hazards Classifications

- Potential injuries include chemical burns from skin contact with hydrochloric acid and burns from hot objects and Bunsen flame.

Laboratory Safety Quiz

1. What should you do if HCl comes in contact with your clothes?
2. Why is it important to keep hair tied back during this experiment?
3. What safety equipment is present in the lab for accidental fires? Where is this equipment located and how do you use it?

Soil Conservation Administration

Proficiency Testing Program

Procedures To Be Tested

Part A. Moisture

1. Obtain and weigh a clean and dry evaporating dish for each soil sample to be tested. Label each dish with the code for the soil to be placed in the dish. Weigh each dish to the nearest 0.01 g. Record the weights. Add the soil sample corresponding to the label to each dish so that the weight of the dish increases by approximately 2 g. Record the weights again. Place the dishes in a drying oven set at 105°C for 24 h.

2. After the 24 h have elapsed, take the evaporating dishes out of the oven (use tongs) and set them on the lab bench for 10 min. Weigh each dish again. Calculate the % moisture in the samples.

Part B. Sample Preparation

1. Dry 25 g of each soil for 24 h in a drying oven set at 105°C.

2. Crush the samples with a mortar and pestle. If a 2-mm sieve is available, pass each sample through the sieve so that the samples to be tested have a maximum size of 2-mm diameter. Store dried and sieved samples in a labeled plastic bag or other suitable container.

Part C. Carbonate Spot Test

1. Place approximately 1 g (by estimation) of each of the dried, sieved soils in separate wells of a porcelain spot plate. Add a small amount of distilled water and stir to remove entrapped air. Then add several drops of 1 M HCl to each and look for any effervescence. Record the effervescence as "none," "trace," "weak," "medium," or "strong." Do not mistake entrapped air bubbles for a positive test.

Part D. Quantitative Carbonate Test

1. For each soil sample containing carbonate (Part C), obtain a clean, dry 125-mL Erlenmeyer flask. Label each flask with the code of the soil for which it is to be used and then weigh each flask. Place 5 g of appropriate soil into the corresponding flask and weigh again so that you can precisely calculate the weight of the soil by subtracting the weight of the empty flask. Wash down the walls of each flask with some distilled water from a squeeze bottle so that all the soil is on the bottom of the flask.

2. Prepare as many small glass vials as you have flasks in the following way. Label each with a code that matches a code on the soil samples. Place 7 mL of 6 M HCl into each vial. Place

a flask (with a soil sample) and the corresponding vial alongside each other on the balance pan to obtain the combined weight to the nearest 0.01 g.

3. Add the acid to the flask so that the acid reacts with the carbonate in the soil. Keep the flask and vial together. Swirl or shake the flask periodically over a period of 30 min. After 30 min, again weigh the flask together with the corresponding vial. The weight loss is the carbon dioxide from the reaction of the acid with the carbonate in the soil. Calculate the % calcium carbonate in the soil.

Part E. Soil pH

1. Perform the following in triplicate on samples of the soil that have *not* been dried. Weigh 5 g of each soil into labeled 50-mL Erlenmeyer flasks or other container suitable for use with the available shaker. Add 25 mL of distilled water. Place on the shaker and shake for 1 h. If there is no shaker available, use a magnetic stirrer set for vigorous stirring.

2. Standardize a pH meter following the instructions for the available pH meter.

3. Let the soil settle for a few minutes after shaking and measure the pH of the solution above the soil. Record all data in an organized manner in your notebook.

Part F. Mineral Content

1. Label three porcelain crucibles and lids with heat-stable labels corresponding to the soil labels. Heat these crucibles with lids ajar with the full heat of a Bunsen burner (so that the bottom of each crucible turns a dull red) for 15 min. Allow to cool for 20 min. Weigh the crucibles to the nearest 0.1 mg and record the weights.

2. Place about 5 g of dried soil in the crucibles and weigh again to the nearest 0.1 mg so that you can determine the weight of the soil.

3. Heat with the Bunsen burner as before, with lids ajar, for 30 min. Cool for 20 min and weigh again. Then, heat for another 15 min. Allow to cool and weigh again. If the crucible lost additional weight, heat for another 15 min, cool, and weigh again. Repeat this until the crucible is at constant weight (no weight changes greater than 0.005 g).

4. The residue remaining in the crucible is the mineral content of the soil. Calculate the % mineral content by dividing the weight of the residue by the weight of the original sample.

Reference

Kenkel J., Kelter P., and Hage D., *Chemistry: An Industry-Based Introduction*, CRC Press/Lewis Publishers, Boca Raton, FL, 2000, Chaps. 7, 12.

Experiment 22
Identifying the
Waste Acid

I.O.N.S.
Innovative Options and New Solutions

Interoffice Memo From Claire Hemistry, CEO

The Klein Chemical Company has discovered an old unlabeled drum of acid in its laboratory's chemical storage area. The company suspects that the label simply fell off as the glue decomposed over time in a rather humid environment and that unsuspecting custodians discarded it during one of their routine cleaning operations. The only label still on the container was one that indicated that the concentration is 5.0 M. Klein would like to dispose of this acid but would like to first identify it. The company has come to us for help.

We are told that it is either sulfuric acid or phosphoric acid, since these are the only acids that the company stores in the area in question. Our contract with Klein calls for us to identify the acid so that they can properly dispose of it. They have sent us a sample of the acid.

Reed P. Ayche of the federal Waste Control Agency is suggesting the laboratory procedure for this work. His memo and procedure are attached. We have also asked Professor Myra Zults of the Waste Management Department at New Trall University to comment on Ayche's procedure. When you are confident that you have correctly identified the acid, please prepare your report memo for the environmental manager at Klein Chemical Company, Bernard Labelov.

Claire

C. Hemistry

aste

ontrol

An Agency of the Federal Government

Dear I.O.N.S. Staff:

For this project, I suggest a titration experiment in which a sample of the acid is gradually neutralized with a base as the pH of the reaction medium is monitored with a pH meter. In case you haven't done a titration experiment before, let me describe it to you. It utilizes a 50-mL buret, which is a long, narrow graduated cylinder with a stopcock valve on one end. I am sure your supervisor can locate one in your laboratory. It also will utilize a pH meter and a combination pH electrode. Basically what you will do is add a solution of the sodium hydroxide from the buret to a solution of the acid held in a 250-mL beaker. The acid will react with the base in what is called a neutralization reaction. The pH of the solution in the beaker will change as the base is added because the acid is being consumed. The addition should take place in increments so that the pH can be monitored properly. For convenience, I suggest setting up an apparatus such that the solution in the beaker can be stirred with a magnetic stirrer. The apparatus will consist of a magnetic stirrer with beaker, a pH meter with combination pH electrode placed in the beaker, and a ring stand with buret mounted so that the tip is just inside the mouth of the beaker (Figure 4.5). A magnetic stirring bar also should be in the beaker. As the solution is stirred, you can add the sodium hydroxide solution in increments (see my procedure for the suggested size of the increments), measuring the pH at the end of each addition.

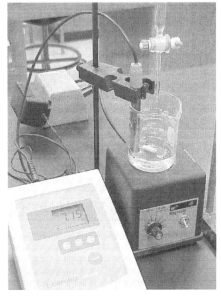

FIGURE 4.5
A photograph showing the complete apparatus for the titration.

    The data obtained will allow the plotting of what is called a "titration curve," a plot of pH on the y-axis and the volume of NaOH added (in mL) on the x-axis. There are some unique features of this curve for each of the two acids in question and these will be the basis for the identification. What I suggest you do is create such a curve for both sulfuric acid and phosphoric acid. You can then create the same curve for the unknown waste acid and then match the curve for the unknown to that of one of the two acids and thereby identify it. Additional details are given in my attached procedure.

Reed

Reed P. Ayche

New Trall YOU

Partners With You in YOUR Education

NEW TRALL UNIVERSITY

Dear Claire,

The titration curves suggested by Ayche will definitely differentiate between sulfuric acid (H_2SO_4) and phosphoric acid (H_3PO_4). The reason has to do with the different number of hydrogens these two acids have and also the relative ease with which these hydrogens are released to the base in the neutralization reaction that is taking place during the titration. Both of sulfuric acid's hydrogens are released to the base readily and the pH will not change appreciably until both hydrogens are neutralized. The reaction is represented by the following chemical equation.

$$H_2SO_4 + 2NaOH \rightarrow Na_2SO_4 + 2H_2O$$

This results in only one discernable sharp increase in pH (what we chemists call an inflection point) during the titration. You will discover this when you graph the results for sulfuric acid.

Phosphoric acid, on the other hand, has three hydrogens per molecule, each of which vary in the ease with which they are released to the base. Because of this, three separate reactions are noticed, one for each hydrogen released, as in the following series of reactions.

$$H_3PO_4 + NaOH \rightarrow NaH_2PO_4 + H_2O$$

$$NaH_2PO_4 + NaOH \rightarrow Na_2HPO_4 + H_2O$$

$$Na_2HPO_4 + NaOH \rightarrow Na_3PO_4 + H_2O$$

This results in three inflection points, two of which will be observable before the final pH = 11 is reached. Thus, the titration curves will be clearly different and the unknown acid easily identified. Good Luck.

Myra

Professor Myra Zults

I.O.N.S. Safety Report
prepared by Ben Whell, I.O.N.S. Safety Coordinator

Experiment 22

Equipment and Technique

- Wear latex gloves to avoid contact with chemicals and solutions.
- Wash hands after handling.
- Always add acid to water, not water to acid, when preparing solutions.
- Use fume hood when preparing acid solutions.
- Keep chemical containers closed when not in use.
- Do not stand on chair or stool to fill buret.
- Do not use a pH meter or stirrer with a frayed cord.

Chemicals

- Sulfuric acid is a dangerous acid. In the concentrated form, it is corrosive and can cause burns to body tissue. If skin contact occurs, rinse immediately. The 0.050 M solution is much less hazardous.
- Phosphoric acid is a dangerous acid. In the concentrated form, it is corrosive and can cause burns to body tissue. If skin contact occurs, rinse immediately. The 0.050 M solution is much less hazardous.
- Sodium hydroxide is a strong base. It can cause irritation to skin, eyes, and respiratory tract and gastrointestinal tract. If skin contact occurs, rinse immediately.

Workplace Cleanup

- Wash hands with soap and water when finished.
- Rinse glassware with water before storage.
- Discard used gloves.
- Dispose of neutralized solutions according to local codes.
- Unused solutions may be labeled and stored.

Hazards Classifications

- Potential serious injuries can result from bodily contact with concentrated acids or solid NaOH. If safety measures are followed, potential injuries are minor and locally treatable.

Laboratory Safety Quiz

1. How should you fill the buret if the top of the buret is out of arm's reach?
2. Why is it important to keep the lid on the bottle of sulfuric acid when not in use?
3. How would you handle a serious spill of concentrated sulfuric acid?

An Agency of the Federal Government

Titration Procedure from Reed P. Ayche

1. Prepare 100 mL of a 0.050 M solution of the waste acid from the 5.0 M solution obtained from Klein Chemicals. If your laboratory doesn't already have 0.050 M solutions of sulfuric acid and phosphoric acid, also prepare 100 mL of each of these from the stock solutions that you have. You will do the titration experiment three times, once with each of these three solutions. If your laboratory doesn't already have a 0.10 M solution of sodium hydroxide prepared, also prepare at least 250 mL of this solution. None of these need to be prepared with any great precision.

2. Set up the apparatus as suggested in my memo attached to this procedure.

3. Standardize the pH meter according to the procedure written for the particular meter you are using.

4. Create a table in the data section of your notebook for the first acid to be tested. The table should be a two-column table with one column headed "pH" and the other headed "mL NaOH added."

5. Start with one of the known acids. Using a graduated cylinder, add 25 mL of the 0.05 M acid to the 250-mL beaker. Place the tip of the combination pH electrode into the beaker such that it is out of the way of the stirring bar. You may need to add some distilled water in order to be sure that the contact to the inside of the electrode is immersed. Start the stirrer. Measure the pH of the solution in the beaker. Record it in your table as the pH for 0.00 mL NaOH added. Fill the buret with the NaOH solution such that the bottom of the meniscus rests on the 0.00 mL graduation line or, if the meniscus is below the 0.00 mL line, record the reading to the nearest 0.01 mL.

6. Be sure to read the buret to the nearest 0.01 mL in all parts of this step. The pH of the acid will not change appreciably for either of the acids up to at least 10.00 mL added. Begin by adding 4.00 mL of NaOH from the buret. Allow sufficient time for complete mixing before recording the pH in your table. Repeat for another 4.00 mL of NaOH (total of 8.00 mL). Then repeat for another 2.00 mL (total of 10.00 mL). You should now begin to add the base more slowly (perhaps about 0.20 mL at a time) and watch the pH carefully to see if it changes appreciably. If it does begin to change, record the pH and the total mL added in your table. If it does not, continue adding about 0.20 mL at a time until it does. The expected change may occur between 10 and 15 mL added, or it may not occur until between 20 and 30 mL, but you need to be ready and add only a drop at a time if the pH is changing rapidly. After this sharp change has occurred, the pH will change more slowly again. You can add the base in larger increments again until another such change is experienced, in which case you should once again add small increments and record the pH and mL of NaOH as you do so.

7.   When the pH reaches 11.0, stop the titration, discard the solution in the beaker (being careful not to lose the stirring bar), and rinse the beaker and stirring bar with distilled water. Place the stirring bar back in the beaker and repeat Steps 5 and 6 for the other acid and then for the unknown acid. When finished, you will have three data tables in your notebook. Make sure that you know which table is for which acid. Label the tables.

8.   Plot the data in the tables to create the three titration curves. The pH should be on the y-axis and the total mL of NaOH added should be on the x-axis. These plots are called titration curves. The identity of the waste acid should be obvious when comparing the three curves. It would be good to include the three curves in your report to Klein Chemicals because Klein will be able to see immediately which is the waste acid.

Reference

Kenkel J., Kelter P., and Hage D., *Chemistry: An Industry-Based Introduction*, CRC Press/Lewis Publishers, Boca Raton, FL, 2000, Chap. 12.

Experiment 23
The Production
Advantage

I.O.N.S.

Innovative Options and New Solutions

Interoffice Memo from Claire Hemistry, CEO

Dear I.O.N.S. Staff:

Please read the attached letter from Speedy Reaction Chemical Manufacturing Company before continuing.

In any chemical reaction, the rate of the reaction can be varied by changing the reaction conditions, such as the concentration of reactants, the temperature, or even the surface area of any solid reactants. In this project, we will address the questions presented to us by the client by reacting magnesium metal with hydrochloric acid under several different sets of reaction conditions and then determine the rate of the reaction by measuring the production of hydrogen gas vs. time.

Since the Speedy Reaction Chemical Manufacturing Company wants to know how the rate of the reaction is affected by (1) the concentration of one reactant (the acid), (2) the temperature of the system, and (3) the surface area of magnesium, we will set up the experiment so as to determine how changes in these variables will affect the rate.

The apparatus that we will use was suggested by our friend, Karen Wosczyna-Birch (see Karen's attached memo and her procedure). It consists of an 8-oz baby bottle and a small plastic bucket. Additional supplies will include the chemicals for the reaction and a timer.

The data will be analyzed graphically by plotting the volume of hydrogen gas produced vs. time for each set of conditions. By comparing the slopes of each graph, we will be able to determine how changing the variables affect the rate of the chemical reaction.

Please address your report memo to R. U. Slower of the Speedy Reaction Chemical Manufacturing Company.

Sincerely,

Claire

Claire Hemistry

Speedy Reaction
Chemical Manufacturing Company

"Winning the Chemical Production Race"

To: Claire Hemistry, I.O.N.S. Corporation

In the business world it is often said that time is money. As CEO of the Speedy Reaction Chemical Manufacturing Company, I am interested in analyzing the factors that could assist us with speeding up chemical reactions so that we can produce more product in less time. We feel that if we optimize the conditions of our process, we would gain a production advantage that would give us a greater share of the business.

Out current focus is the production of hydrogen gas by reacting hydrochloric acid and magnesium metal. The reaction that we are using is as follows:

$$Mg(s) + 2HCl(aq) \rightarrow MgCl_2(aq) + H_2\uparrow$$

We ask that you research various reaction conditions to determine which will provide for the greatest amount of hydrogen in the least amount of time. The conditions we would like you to explore are (1) concentration of the acid, (2) temperature of the system, and (3) the surface area of the magnesium.

I very much look forward to the results of this study.

Sincerely,

R. U. Slower

R. U. Slower, CEO

From the office of Karen Wosczyna-Birch

Dear I.O.N.S. Staff:

I am happy to help with this project. I think the use of the baby bottle to contain the reaction will help because the production of hydrogen gas will result in the acid solution being forced out through the nipple and so the solution volume in the bottle will decrease with time. The amount of the decrease is equal to the volume of hydrogen produced at any given point in time, so how this volume changes with time will be an indication of the reaction rate.

The supplies consist of an 8-oz baby bottle, various forms of magnesium (ribbon, turnings, and powder), a 1-gal plastic bucket, and various concentrations of hydrochloric acid. The idea is to perform the reaction a number of times corresponding to the different reaction conditions. So, while working with magnesium ribbon and room temperature, you can change the concentration of the HCl. Then, with a given concentration of HCl and at room temperature, you can change the form of the magnesium. And finally, while using one form of magnesium and one concentration of HCl, you can vary the temperature.

To be most efficient, I would suggest that at least three groups be formed, each to investigate a particular variation. One group can investigate the effect of concentration, another group investigates the effect of temperature, and a third studies the effect of the different forms of magnesium. Data from all groups will then be shared.

My procedure is attached. Good Luck.

Sincerely,

Karen

Karen Wosczyna-Birch

I.O.N.S. Safety Report
prepared by Ben Whell, I.O.N.S. Safety Coordinator

Experiment 23

Equipment and Technique

- Wear latex gloves and handle HCl solutions with care. A fume hood is recommended.
- Be cautious with magnesium turnings and powder so that you don't breathe in any airborne particles.
- Wear latex gloves when handling the magnesium ribbon.
- Although it is a low concentration, a fume hood is recommended for the warming of the 0.5 M HCl solution.
- Do not use a hot plate with a frayed cord.
- Hydrogen gas is highly flammable. Avoid flames and sparks.

Chemicals

- The 2.0 M HCl is somewhat dilute, but still a potentially dangerous acid solution. Neutralize spills with a solution of a weak base.
- Magnesium metal is safe to handle, but latex gloves are recommended.

Workplace Cleanup

- Spent acid solutions may still be acidic. They also contain dissolved magnesium. Flush down the drain with plenty of water.
- Unused portions of acid solutions may be stored in stoppered, labeled containers.
- Baby bottles and glassware may be rinsed and stored.

Hazards Classifications

- Potential injuries are minor and treatable on site.

Laboratory Safety Quiz

1. Why are diluted acid solutions less hazardous than concentrated acid solutions?

Karen's Procedure

Preliminary Note

Check out the bottle/bucket arrangement and obtain the "time zero" reading by performing the following preliminary test. Nearly fill the bucket with water. Place 200 mL of water in the bottle. Use the milliliter scale on the side of the bottle for this measurement. The bucket should be big enough and have enough water in it so that when the bottle, containing 200 mL of water, is inverted into the bucket, it will float. Thus, with the inside and outside liquid levels the same, no liquid should leak through the nipple of the bottle until the hydrogen gas is generated. It must also be possible to move the floating bottle to a vertical position so that the volume of the liquid inside the bottle can be accurately read using the milliliter scale on the side of the bottle. Record this volume reading now and use this as the "time zero" reading for all trials. Empty the bottle.

Summary Procedure

Place between 0.10 and 0.15 g of magnesium in the nipple of the baby bottle (see below for the technique for different forms). Place 200 mL of the acid solution in the bottle. Carefully and tightly screw the nipple cap on the bottle such that the no magnesium falls out prematurely. Then simultaneously invert the bottle into the bucket and start the timer. The reaction should start immediately, releasing the hydrogen into the bottle. Read the volume of solution, using the milliliter scale on the side of the bottle, every 30 sec until the magnesium is used up and the reaction stops. Discard the solution.

Varying the HCl Concentration

Prepare 250 mL of each of three solutions of hydrochloric acid with concentrations varying between 0.2 and 1.0 M. Use a stock solution that is 2.0 M for the preparation. Perform the **summary procedure** three times, once for each concentration, at room temperature using the form of magnesium of your choice (same form each time).

Varying the Form of Magnesium

Prepare 700 mL of 0.50 M HCl from the stock solution that is 2.0 M and use this acid for the three tests with the different forms of magnesium. If the magnesium is in ribbon form, fold it in an accordion-like fashion and push it into the nipple of the baby bottle. If the magnesium consists of turnings or powder, you may "pour" it into the nipple from the weighing paper, but be careful not to spill it into the acid before you are ready. Perform the **summary procedure** for each form.

Varying the Temperature

Prepare 800 mL of 0.50 M HCl. Perform the **summary procedure** with a 200 mL portion of it and the form of magnesium of your choice. Then, heat 250 mL of this solution to 30 to 35°C on a hot plate. Also heat the water to be used in the bucket to the same temperature on another hot plate. Place the warm water in the bucket. Perform the **summary procedure** using the warm acid and the same form of magnesium as in your first trial. Measure the temperature of the warm water in the bucket while the experiment is running and report this as the temperature of the test.

Prepare cold water by mixing distilled water with ice in the bucket. Use this cold water to prepare 250 mL of 0.50 M HCl. You may have to add 250 mL of water back to the bucket. Perform the **summary procedure**. Measure the temperature of the water in the bucket about midway through the test and report this as the temperature of the test.

Each group will now **prepare three graphs** as follows:

> **Effect of acid concentration:** Plot the volume of hydrogen gas produced vs. time for each concentration.
>
> **Effect of different forms of magnesium:** Plot the volume of hydrogen gas produced vs. time for each form used.
>
> **Effect of temperature:** Plot the volume of hydrogen gas produced vs. time for each temperature.

With the data from all groups available, you should be able to conclude what set of conditions will work best for Speedy Reaction Chemical Company. You can explain it to them by noting the slope of each graph.

Acknowledgment

This experiment was suggested by Karen Wosczyna-Birch, Tunxis Community College, Hartford, CT.

Reference

Kenkel J., Kelter P., and Hage D., *Chemistry: An Industry-Based Introduction*, CRC Press/Lewis Publishers, Boca Raton, FL, 2000, Chap. 14.

Experiment 24
Does It or Doesn't It: The
Fruit Juice Controversy

I.O.N.S.
Innovative Options and New Solutions

Interoffice Memo from Claire Hemistry, CEO

Dear Staff:

We have been approached by Nectar of the Gods Fruit Juice Company for assistance with the following problem. As a part of their advertising, they claim that their fruit juice products contain all nine essential amino acids. A consumer group has recently challenged that claim. The consumer group has boldly claimed that one of the juices totally lacks four of the essential amino acids: leucine, methionine, valine, and lysine. We have been retained by Nectar of the Gods to verify, via laboratory analysis, that all four of these amino acids are present in this product. The amounts of each is not our concern. We simply have been asked to determine, yes or no, if they are present. Our analyses, therefore, will be qualitative, but not quantitative.

Our consultant for this project is Dr. Lou Seen, an expert in the laboratory analysis of amino acids. I've attached his comments and recommendations to this memo. I'm sure you will find his advice useful. Good luck on this project. The *Manual of Food Analysis* procedure directs you to report to your "client or customer." However, please address your report to me and I will prepare the formal report for Nectar of the Gods.

I also noticed that the *Manual of Food Analysis* suggests that you wear safety glasses when carrying out the procedure. I don't think I have to remind I.O.N.S. laboratory technicians that safety glasses are to be worn in our laboratories at *all* times.

Sincerely,

Claire

C. Hemistry

From the desk of Dr. Lou Seen

August 13, 1999

Claire,

I believe that Nectar of the Gods' problem may be solved fairly easily in the laboratory. Both thin-layer chromatography (TLC) and paper chromatography are effective methods for separating and identifying amino acids. These techniques are both qualitative and, thus, should work well for this analysis. All you need, in addition to the fruit juice in question, are pure solutions of the four amino acids in a solvent such as 70% isopropyl alcohol. Ascending chromatography should separate the amino acids. As a mobile phase (I recommend 70% isopropyl alcohol) ascends the paper or silica gel TLC plate, the amino acids also will move, but at different rates. By comparing the fruit juice with the known amino acids, you should be able to verify the presence of all four. Details of this procedure may be found in the procedures in the *Manual of Food Analysis,* published by the Federal Agency of Food Nutrition.

I also recommend using both TLC (silica gel) and paper chromatography. This will add additional credibility to your results.

Please feel free to contact me if you require additional assistance.

Lou

I.O.N.S. Safety Report
prepared by Ben Whell, I.O.N.S. Safety Coordinator

Experiment 24

Equipment and Technique

- Handle glass jar and lid carefully to avoid breakage.
- Use caution when spraying with the ninhydrin solution to avoid getting it on skin or clothing.
- Use caution handling hot items from drying oven.

Chemicals

- Isopropyl alcohol is poisonous and flammable. Do not ingest and keep away from flames.
- The amino acids used in this experiment pose no major, significant safety problem.
- Ninhydrin is the hydrate of indane-1,2,3-trione. It may cause skin irritation and discolor the skin. Use latex gloves when preparing the solution and avoid getting the spray solution on the skin.

Workplace Cleanup

- The used paper and TLC plates may be discarded in the trash.
- The used capillary tubes can be discarded in the trash receptacle designated for broken glassware.

Hazards Classifications

- Potential injuries are minor and treatable on site.

Laboratory Safety Quiz

1. What specific safety hazards does isopropyl alcohol present?

Manual of Food Analysis

Procedure 112. Chromatographic Separation of Amino Acids

Materials and Equipment Required

1. Whatman #1 chromatography or filter paper: 25×16 cm
2. TLC plates (silica gel): 25×16 cm
3. 0.01 M solutions of leucine, methionine, valine, and lysine in 70% isopropyl alcohol
4. Food sample in liquid form
5. Mobile phase, 70% isopropyl alcohol
6. Capillary tubes (melting point capillaries open on both ends)
7. Paper clips
8. Covered developing jar
9. Ninhydrin solution (0.1% in butanol or isopropyl alcohol solvent) in spray bottle
10. Drying oven
11. Ruler

Experimental Procedure

Note: Safety glasses should be worn throughout this procedure.

1. Pour mobile phase into the developing jar to a depth of 2.0 cm. Cover the jar.
2. Draw a horizontal line (**in pencil**) across the paper and TLC plates at a distance of 2.5 cm from one edge as shown in Figure 4.6. Draw short vertical lines at 3.5 cm intervals across the horizontal line as shown. The intersections of these lines are where sample spots will be applied. Also, label each intersection as shown. **IMPORTANT** — Since amino acids are found on human hands, do not touch the paper or plate. Handle only by the edges. Also, lay the paper and plate on paper towels while setting up. For marking the paper or plate, **use pencils only**.
3. Using separate capillary tubes for each sample, spot the paper or plate by placing a *small* drop from the capillary tube at the appropriate points (where the lines cross). The smaller the spot, the better. *Perform this step carefully.* Allow the spots to dry. Sample spots, because the samples may be dilute, should be spotted a second time over the original spot.
4. When all samples are spotted, roll the paper carefully, keeping the bottom even, paper clip the top, and place into a developing jar. The plates are placed directly in the jar.
5. After 2 hours *minimum* (3 is better) remove the paper and plate, *quickly* mark the solvent front (the edge of the dampness) in pencil, and place in a fume hood to dry.
6. After drying 5 to 10 minutes, spray the developed portion with the ninhydrin solution. Allow to dry 10 minutes.
7. Reclip paper and place it and the plate in a drying oven at 60°C for 5 minutes. Spots representing amino acids will be violet to blue.

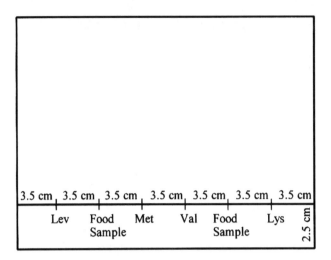

FIGURE 4.6
A drawing of the paper or plate showing the positions of the spots and the distances involved.

Interpretation of Results

The purpose for performing this procedure is to determine which amino acids, if any, are found in the food sample. This may be ascertained by simply comparing the known amino acid spots with the spots from the food sample. For example, if valine migrates 4.0 cm from the origin and the food sample shows an amino acid spot 4.0 cm from the origin, the logical conclusion is that the spot **is** valine. A more mathematical approach involves calculating retardation factors (R_f) values for all spots, with R_f defined as:

$$\frac{\text{Distance Spot Moved}}{\text{Distance Solvent Moved}}$$

To calculate R_f values, first circle all spots and mark the center of each. Then measure the distance from the origin to the center of the spot and divide that number by the distance from the origin to the solvent front.

Final Report

Write a memo to your client or customer detailing your findings from this lab procedure. Document your results as thoroughly and carefully as possible. Remember that your company's reputation is on the line!

Acknowledgment

This experiment was written by Don Mumm, Southeast Community College, Lincoln, NE.

Reference

Kenkel J., Kelter P., and Hage D., *Chemistry: An Industry-Based Introduction*, CRC Press/Lewis Publishers, Boca Raton, FL, 2000, Chap. 17.

Experiment 25
The Fun Fat
Analysis

I.O.N.S.
Innovative Options and New Solutions

Interoffice Memo from Claire Hemistry, CEO

Dear Staff:

The Jupiter Chocolate Company is a well-known maker of candy bars. The content of fat in these bars is critical. Company specifications call for not less than 30% and not more than 40% fat in each candy bar. Jupiter has contracted with us to monitor fat content in their products. Our task will be to quantitatively determine the % fat in a variety of their bars and report back to Jupiter with our results.

Professor Li Pidd at State University is our consultant on this project. I've attached his recommendations to this memo. Good luck on this project! I look forward to reading your final results and report.

Sincerely,

Claire

C. Hemistry

P.S. I see that the *Manual of Food Analysis* once again provides a special safety guideline here. Remember, I.O.N.S. safety policies requires the wearing of safety glasses at *all* times.

State University

From the desk of Professor Li Pidd

Dear Claire:

The ability to determine fat content in foods is vital. Fortunately, it is also routine. Since fats are insoluble in water (hydrophobic, which literally means "water fearing"), they may be separated fairly easily from the other components of a food sample via a simple extraction with a hydrophobic solvent such as ether. The procedure generates very accurate data. A word of caution, however. Ether is highly flammable and volatile and, therefore, must be handled with extreme caution to avoid contact with open flames. For procedural details, refer to Procedure #65, "Fat Extraction and Determination" in the *Manual of Food Analysis*.

If your technicians have any problems with this procedure, I am available to assist them.

Li

I.O.N.S. Safety Report
prepared by Ben Whell, I.O.N.S. Safety Coordinator

Experiment 25

Equipment and Technique

- A round-bottom flask will not stand upright by itself. Use a 250-mL beaker to hold the flask so it doesn't roll off the benchtop and break.
- Use boiling chips in the water bath so the water does not superheat causing "bumping" and possible breakage.
- Be sure to use a fume hood to vent the hydrochloric acid fumes and ether fumes.
- Use insulating gloves to handle hot glassware.
- When extracting the fat with the ethers, remember to release the pressure in the round-bottom flask frequently, as directed.
- Use the fume hood and NO FLAMES OR SPARKS in the lab while performing this experiment.
- Do not use a hot plate with a frayed cord.

Chemicals

- Concentrated hydrochloric acid is a strong acid with stifling fumes. Be very careful to prevent accidental breakage and spills when transferring the 10 mL and handling the round-bottom flask. Hot, concentrated hydrochloric acid is especially dangerous. Use only in a good fume hood.
- Ethyl ether is a highly volatile and highly flammable liquid. It evaporates readily at room temperature and the dangerous vapors fill a room quickly. It must be confined to the fume hood as much as possible.
- Petroleum ether is a mixture of volatile and flammable hydrocarbons. Like ether, the dangerous vapors fill a room quickly. It, too, must be confined to the fume hood as much as possible.
- If ether fumes become strong, e.g., due to a spill outside the fume hood, vacate the lab until the vapors subside.
- Ethyl ether slowly decomposes to form explosive peroxides over time. Store in original metal can in a refrigerator, preferable one that is explosion-proof. Do not store in the lab more than 9 months.

Workplace Cleanup

- Dispose of fat residues only as directed by your supervisor.

Hazards Classifications

- Potential injuries are minor and treatable on site. If fumes cause lightheadedness, seek fresh air immediately.

Laboratory Safety Quiz

1. What would you do if you noticed that a full can of ethyl ether has been in the lab more than a year?
2. What would you do if the fume hood suddenly stops due to a tripped circuit breaker?

Manual of Food Analysis

Procedure 65: Fat Extraction and Determination

Materials and Equipment Required

1. 100-mL round-bottom extraction flasks with stoppers
2. Concentrated hydrochloric acid (HCl)
3. Analytical balance
4. Water bath: 250-mL pyrex beakers half-full of tap water on a hot plate
5. Ethyl ether
6. Petroleum ether
7. Freezer
8. 600-mL beaker

Experimental Procedure

Note: Safety glasses and rubber gloves must be worn throughout this procedure. No open flames.

1. Using an analytical balance, weigh about 1 g of the food sample on a piece of weighing paper. Record this weight to the nearest 0.01 mg.
2. Place a 100-mL round-bottom extraction flask in 250-mL beaker to hold it upright. Add the sample to the flask. Label the flask and the beaker with your initials.
3. Place the flask, without a stopper, in the water bath. The water bath must be under a fume hood. Label the water bath with your initials.
4. Add 10.0 mL of concentrated hydrochloric acid (*Caution* — strong acid) to the flask.
5. Heat for 45 min at a gentle boil. Swirl the flask at 10 min intervals.
6. After 45 min, remove the flask from the water bath and add 10 mL of distilled water. (This is safe to do because the hydrochloric acid strength has greatly diminished due to the reaction that has occurred.)
7. Cool the flask by running cold water over the exterior of the flask.
8. Stopper the flask, place it back in the 250-mL beaker (to hold it upright) and place it in a freezer for 30 min. This aids the fat separation.
9. After 30 min, remove flask from the freezer and, while working in a fume hood, add 25 mL of ethyl ether.
10. Continuing to work in a fume hood, stopper the flask and shake vigorously for 30 sec, releasing pressure every 10 sec by removing the stopper.
11. Add 25 mL of petroleum ether to the round-bottom flask and repeat the extraction procedure, again releasing pressure every 10 sec.
12. Carefully decant the ether (top) layer into a tared (preweighed) 600-mL beaker.
13. In the fume hood, set the beaker on a hot plate (lowest setting) to evaporate the ether.
14. Repeat the extraction with both ethers two more times, decanting into the beaker each time.
15. Leave the beaker on the hot plate until all of the ether is evaporated, leaving only the fat.
16. Cool and weigh the beaker plus the fat and record the weight.

Interpretation of Results

The combination of the two ethers and the repetitive extractions will easily extract the fat from the food sample. All other food components will remain in the bottom, aqueous layer. The HCl plus heat combination breaks up the food sample, releasing free fatty acids, monosaccharides, amino acids, etc., thus making the extraction easier. Calculate % fat in the sample as follows:

$$\% \text{ fat } = \frac{\text{Weight of fat}}{\text{Weight of sample}} \times 100$$

The weight of fat is determined by subtracting tared beaker weight from weight of beaker plus fat.

Final Report

Write a memo to your client or customer listing all samples with corresponding fat percentages of each. Also indicate whether or not the numbers fall within company specifications.

Acknowledgment

This experiment was written by Don Mumm, Southeast Community College, Lincoln, NE.

Reference

Kenkel J., Kelter P., and Hage D., *Chemistry: An Industry-Based Introduction*, CRC Press/Lewis Publishers, Boca Raton, FL, 2000, Chap. 17.

Appendix

Author's Statements

Supplies and Equipment Lists

Experiment 1. Basics of Weight and Volume Measurement

Author's Statement

This experiment is designed to introduce students to some simple measuring devices in the laboratory so that when they are used in subsequent lab activities, students will know how to use them properly and will have some idea of their precision.

Equipment

Variety of balances that can be read the nearest to 0.1 g, 0.01 g, and 0.0001 g

Small beakers

10-mL graduated cylinders

50-mL burets

Various other devices intended to measure 4 mL of water to different degrees of precision.

  a. Burets: 10 mL, 250 mL
  b. Pipetters
  c. Variety of droppers (glass and plastic)
  d. Plastic squeeze bottles for dispensing drops of water

Experiment 2. Identifying Ordinary Household Products

Author's Statement

This experiment is designed to introduce chemical and physical properties and how these properties can be used for identification purposes. Since Part A is done as a group, it also promotes teamwork. This could be a 2-week activity if the instructor chooses — Part A the first week and Part B the next.

Equipment

Small glass vials to contain white solids, knowns and unknowns

Small test tubes

Test tube racks

Labeling tape

Spatulas

10-mL graduated cylinders

Squeeze bottles for distilled water

Plastic dropper bottles to contain reagents

Porcelain spot plates

Hot water bath (hot plate with beaker)

Reagents

Tincture of iodine

Vinegar

Phenolphthalein indicator solution

0.3 M NaOH

Rubbing alcohol

Benedict's Reagent

White Solids

Table salt

Table sugar

Fruit sugar (fructose — can purchase at healthfood store)

Drain opener (solid, 100% sodium hydroxide, such as Red Devil Lye)

Cornstarch

Baking soda

Washing soda (available at supermarkets)

Plaster of paris

Epsom salts

Boric acid (available at pharmacies as an insecticide)

Calcium supplement (calcium carbonate available at healthfood stores and pharmacies — do not use antacid tablets)

Experiment 3. Five Vials: Identifying Dissolved Cations and Anions

Author's Statement

Besides chemical and physical properties and how these properties can be used for identification purposes, this experiment helps students learn the formulas and names of monatomic and polyatomic ions and allows them to observe light emission and absorption, which may be part of lecture discussions occurring at the same time. Since Parts A and B are done as a group, it also promotes teamwork. This is intended as a 2-week lab — Parts A and B the first week and Part C the next. With 10 ions to identify, Part C teaches time management.

Equipment

Dropper bottles for reagents and known solutions (also larger bottles for stock solutions)

Small test tubes and test tube racks

10-mL graduated cylinder

Centrifuge

Atomic absorption instrument (or other flame test equipment)

Small watch glasses

Hot water bath (hot plate and beaker)

Disposable polyethylene pipets (such as Fisher #13-711-7)

Apparatus and chemicals for making chlorine water (see below)

    a. 500-mL Florence flask

    b. Thistle tube

    c. Glass tubing

    d. Two-hole rubber stopper for Florence flask

    e. 2-ft length of rubber tubing

    f. 250-mL Erlenmeyer flask with rubber stopper

UV spectrophotometer (with quartz cuvettes)

Vials for unknowns (screw cap, 30 mm)

Racks for vials (Fisher 14-791-5E)

Procedure for Making Chlorine Water

We recommend that the instructor or stockroom worker prepare the chlorine water in advance of the lab session and keep it tightly stoppered. Near the beginning of the lab session, it can be transferred to several plastic dropper bottles with caps tightly screwed.

Set up a 500-mL Florence flask with two-hole stopper as follows. Insert a thistle tube (polypropylene — for safety) into one of the holes in the stopper such that the tubing end almost touches the bottom of the flask when the stopper is fitted into the mouth of the flask. Insert an ordinary glass tube in the other hole. Insert the stopper into the mouth of the flask. Obtain a 2-ft section of rubber tubing. Place one end on the glass tube extending from the two-hole stopper and the other on a piece of glass tubing, perhaps 4 in. long. Place the open end of this latter tube under 200 mL of water held in a 250-mL Erlenmeyer flask. Hold the flask in place on a hot plate using a clamp and a ring stand.

Remove the stopper and place a quantity of manganese dioxide in the Florence flask, enough to cover the bottom of the flask to a depth of about a millimeter. Replace the stopper. Add concentrated hydrochloric acid (20 to 30 mL) through the funnel in the top of the thistle tube so that it soaks the manganese dioxide in the flask and so that the bottom of the thistle tube is below the surface of the acid. Begin heating. Chlorine gas is generated in the flask and bubbles into the water in the Erlenmeyer flask. Control the production of chlorine by controlling the heat. Continue the heating and bubbling until the water in the Erlenmeyer flask is a light yellow color due to the dissolved chlorine.

For cleanup, the material in the Florence flask should be diluted with water and the diluted solution decanted down the drain with plenty of water. The manganese dioxide residue should be collected and stored for proper disposal according to local guidelines.

The tubing and glassware should then be rinsed with copious amounts of water until the chlorine odor has dissipated. The apparatus can be stored for later use.

Reagents

Concentrated HCl	1 M KCNS
6 M HCl	Dimethylglyoxime solution
3 M HCl	(1% weight/volume in acetone or ethanol)
0.1 M HCl	1 M Na_2SO_4
Concentrated HNO_3	Hexane
6 M HNO_3	Ammonium molybdate solution
3 M HNO_3	0.2 M $BaCl_2$
Concentrated NH_4OH	0.1 M $AgNO_3$
6 M NH_4OH	1 M $(NH_4)_2CO_3$
4 M NaOH	Fresh chlorine water
1.5 M H_2SO_4	(as per instructions given here)

Chemicals

Each known and unknown should be a 0.1 M solution of an ionic compound. There are a large number of ionic compounds that could be listed that would be useful. For known solutions of the anions, the sodium salts (NaCl, NaBr, NaI, Na_2CO_3, Na_2SO_4, Na_3PO_4, $NaNO_3$), or any soluble salts containing the anions, would be useful. For the cations, the chloride salts (NaCl, $CaCl_2$, KCl, LiCl, $BaCl_2$, $AlCl_3$, $NiCl_2$, $FeCl_3$, NH_4Cl), or any soluble salts containing the cations, would be useful. For unknown solutions, soluble compounds are chosen so that the desired cation and anion are present. For example, to create an unknown containing Ca^{2+} and NO_3^-, a 0.1 M solution of $Ca(NO_3)_2$ would be used.

Experiment 4. Practice in Naming and Formula Writing

Author's Statement

Lecture discussions and exams and quizzes in which students are asked to assign names to formulas, or assign formulas to names, are often not fun for either students or teachers. This "experiment" can help in any number of ways. It can be a homework (group or individual) exercise to support the lecture discussion. It can be an in-class group exercise. It can be a lab activity (group or individual). At the instructor's discretion, it can either support the lecture discussion or be the only formal exposure to naming and formula writing the students receive while the classroom discussion emphasizes topics more interesting to students and teachers.

Experiment 5. A Chemical Scavenger Hunt

Author's Statement

Identifying consumer products that contain the chemicals students are studying in the classroom adds a touch of realism to what they are learning. The introduction to the

Handbook of Chemistry and Physics, the *Merck Index,* MSDSs, and the Fisher chemical catalog exposes them to some elements of the chemist's workplace that they would not otherwise get. Actually seeing the chemicals that are in ordinary products that they use as consumers, and reading the label on the containers, communicates a sense of the relevancy about the role of chemistry in their lives. Finally, the use of the CD-ROM and the Internet (optional) gives them a first clue to modern technology as a part of the chemist's workplace

Consumer Products (or their Empty Containers Showing the Ingredients List)

Beef jerky (or other product containing sodium nitrite)

Sourdough English muffins (or other product containing ammonium sulfate)

Crest toothpaste (or other product containing sodium fluoride)

Baking soda (or other product containing sodium bicarbonate)

Shampoo (or other product containing potassium chloride)

Liquid drain opener (or other product containing sodium hypochlorite)

Nasal decongestant (or other product containing calcium phosphate dibasic)

Baking powder (or other product containing calcium sulfate)

Boxed macaroni and cheese (or other product containing ferrous sulfate)

Ice melt (or other product containing calcium chloride)

Iodized table salt (or other product containing sodium thiosulfate)

Washing soda (or other product containing sodium carbonate)

Equipment and Materials

Copies of the *Handbook of Chemistry and Physics*

Microcomputers (optional) with CD-ROM drives

Merck Index CD-ROMs (or printed copies of the *Merck Index*)

Microcomputers connected to the Internet (optional) (or other sources of MSDSs)

Original containers of chemicals (with labels giving the formula weights)

Fisher chemical catalogs

Experiment 6. The Laboratory Safety Exercises

Author's Statement

This experiment, if properly executed, follows the guidelines found in an American Chemical Society publication (*Safety in Academic Chemistry Laboratories*) which state that fume hoods, eyewash stations, and safety showers should be tested regularly for proper operation and that students should know how to operate these devices. It also serves to help bring safety issues to the forefront in the students' study of chemistry.

Materials

Book: *Building Student Safety Habits for the Workplace,* published by Terrific Science Press, Miami University Middletown, Middletown, OH, 2000, ISBN 1-883822-18-1

Laboratory fume hood

Velometer

Tissue paper

Tape

Matches

Smoke generator

Tape measure

Commercial safety shower test kit

Bucket

Eyewash gauge

Experiment 7. Out of Spec — Out of Mind

Author's Statement

This experiment is designed to support classroom discussions about light, the electromagnetic spectrum, light absorption, spectrophotometers, etc. It also brings to light a typical problem real-world industrial chemists face at times, and that is the analysis of an off-color or out-of-spec material. This experiment could be combined with Experiment 8 in the same lab period.

Equipment and Materials

50-mL beakers

Top-loading balances

Spatulas

Visible spectrophotometer

Cuvettes

Cotton swabs

Graph paper (or microcomputer with graphing software and printer)

Chemicals

Ferric chloride (contained in small labeled vials)

Nickel chloride (contained in small labeled vials)

Ferric chloride mixed with colored soluble contaminant (in small labeled vials)

Nickel chloride mixed with colored soluble contaminant (in small labeled vials)

Experiment 8. Assuring the Quality of a Copper Reference Standard

Author's Statement

This experiment is designed to support classroom discussions about light, the electromagnetic spectrum, light absorption, spectrophotometers, standard curves, etc. It takes Experiment 7 one step farther in that it demonstrates a quantitative analysis using a spectrophotometer, which involves the creation of the standard curve. The application to a commercial reference standard connects the students to the chemist's workplace in which the chemist is always concerned about the quality of the standard materials that are available. Students may be able to do both Experiment 7 and Experiment 8 in a 3-hour lab period if the instructor chooses. The experiment is meant to be used early in a course and, thus, does not require a full understanding of concepts of solution concentration.

Equipment

Weighing paper

Top-loading balance

Funnels

Volumetric flasks, 100-mL

Graduated cylinders, 10-mL

Labeling tape

Pipetters

10-in. stirring rods flattened on one end (these can be made by heating one end in a flame and pressing vertically on a hard surface)

Visible spectrophotometer

Cuvettes

Graph paper (or microcomputer with graphing software and printer)

Chemicals

$CuSO_4 \cdot 5H_2O$

6 M NH_4OH contained in plastic dropper bottles

Copper reference standard (1000 ppm)

Experiment 9. No Labels. Now What?

Author's Statement

This experiment is designed to be a first experience with common organic liquids, specifically the physical properties of these liquids, and to support classroom discussions about simple organic chemicals as well as density and other physical properties. It also exposes students to what can be an important activity in the chemist's workplace — the identification of unknown or unlabeled chemicals.

Equipment

Latex gloves

Large test tubes with screw caps for containing knowns and unknowns

Disposable polyethylene pipets (such as Fisher #13-711-7)

Small test tubes

Test tube racks

10-mL burets or 10-mL graduated cylinders for density measurement

Top-loading balances

Small beakers

50-mL buret that has been cut off at about the 38-mL line and fire polished.

Stop watch

Refractometer

Experiment 10. The Frustrating Federal Film Folly

Author's Statement

This experiment is meant to accompany classroom discussions about polymers and infrared spectrometry. It connects students' personal life (commonly used plastic films) to the chemist's workplace (analysis procedures designed to characterize materials), which is an ongoing and important objective of the experiments in this book.

Equipment and Materials

Infrared spectrometer

Cardboard mounting brackets

Mounting tape

Plastic film (polystyrene, polyethylene)

Unknowns brought from home

Experiment 11. MSDSs vs. Labels — What's Missing?

Author's Statement

The ability to understand and apply the information contained in MSDSs is very important for chemists in the modern workplace. This exercise highlights that aspect of a chemist's work and also reemphasizes (after having done Experiment 6) the importance of safety considerations in a chemistry laboratory. It also connects chemistry to the students' personal lives by requiring a close look at the chemical information found on the labels on common consumer products.

Materials

Book: *Building Student Safety Habits for the Workplace* published by Terrific Science Press, Miami University Middletown, Middletown, OH, 2000, ISBN 1-883822-18-2

Experiment 12. Mistakes at a Steel Plant

Author's Statement

This experiment is intended to support classroom discussions concerning significant figures, the metric system, and unit conversions. The exercise with galvanized nails is optional. In the absence of the nails, this experiment could be combined with Experiment 13 in one lab period.

Materials

Latex gloves

Galvanized sheet metal (can be purchased inexpensively in rolls at builders supply stores)

Galvanized nails (optional)

Metal shears (to cut out rectangular steel samples that are approximately 3 cm × 12 cm)

Top-loading balance

Rulers calibrated in centimeters

Fume hood

Containers for acid (containers must be deep enough to have 12-cm sample totally immersed)

Tongs or tweezers

Chemicals

6 M HCl (Author comment concerning the acid: One container of acid can be used repeatedly for the purpose, perhaps even be stored from one semester to the next to be used again and again. Eventually the strength diminishes and it will take longer to dissolve the zinc. The acid solution also will become highly colored due to the presence of dissolved zinc and iron. At that point, fresh acid should be prepared and disposal procedures utilized for the used acid.)

Experiment 13. The Case of the Cracked Engine Block

Author's Statement

This experiment is intended to be an exercise in the application of significant figures and the metric system to the measurement of a physical property, density. It is intended to support the classroom discussion of these topics. Waste antifreeze can be disposed of through an automotive shop at a college or through an automotive service station.

Equipment

Latex gloves

10-mL graduated cylinders

100-mL beakers

Droppers

Labeling tape

Large container for waste antifreeze

Graph paper (or computer with graphing software)

Chemicals

Commercial antifreeze

Antifreeze–water mixture that is the unknown to be tested

Experiment 14. Quality Control Problems at a Fluids Plant

Author's Statement

As Experiments 12 and 13, this experiment supports a classroom discussion concerning significant figures, the metric system, and the measurement of physical properties. In addition, it introduces the concept of control charting, a popular and important means of monitoring the quality of measurements and products in quality assurance laboratories. It, thus, utilizes a familiar consumer product to demonstrate an activity of the chemist's real world of work and to tie in the classroom discussion.

Equipment

Latex gloves

Pycnometers (25 or 50 mL)

Top-loading balances

Kimwipes™

Graph paper for control chart (or computer with software for control charting)

Chemicals

Methyl alcohol (for rinsing pycnometers)

Commercial windshield washer fluid

Windshield washer fluid samples held in 10 separate bottles labeled "Day 1," "Day 2," etc. (One or more of the above samples may be purposely contaminated with water. One gallon of fluid will provide 10 samples of 375 mL each.)

Recommendations for Control Chart

The instructor should purchase the windshield washer fluid to be used and then, in advance, carefully measure its specific gravity as will the students. This value can be used as the "desirable value" in the control chart. The warning limits can be set at ±0.002 and the action limits at ±0.003 and the students told that the resulting control chart is what the fluids company has been using. The students then create a control chart, chart their data, and interpret the results.

Experiment 15. Matchmaker's Dilemma

Author's Statement

This experiment supports the classroom discussion of stoichiometric calculations involving chemical equations. It also teaches the correct use of a Bunsen burner. Students may need to be reminded that the SOP is for pure $KClO_3$. They will have to make a minor change in the procedure and calculation to adapt it to the Matchmaker Plus sample and the purpose of the experiment.

Equipment

Bunsen burners

Strikers or other flame lighters

Porcelain crucibles and lids

Ring stands with rings

Clay triangle

Spatulas

Analytical balance

Tongs

Chemicals

Mixture of $KClO_3$ and KCl labeled "Matchmaker Plus." (This sample should either be a little higher than 50% $KClO_3$ or a lower than 50% $KClO_3$, so that students can conclude that it either meets specifications or it doesn't.)

Experiment 16. Salt From Soda Ash — Will It Work?

Author's Statement

The experiment supports the classroom discussion concerning stoichiometric calculations, in particular, % yield calculations. It also promotes critical thinking skills because the student is asked to assess the adaptability of the reaction to full-scale industrial production of NaCl.

Equipment

Porcelain crucibles and lids

Ring stands with rings

Clay triangles

Bunsen burners

Spatulas

Analytical balance

Tongs

Chemicals

Concentrated HCl (in plastic dropper bottles)

Sodium carbonate

Experiment 17. A Proposed New Analytical Method

Author's Statement

This experiment supports classroom discussions concerning stoichiometric calculations involving solutions. It also introduces the students to the use of a spectrophotometer to measure turbidity. The "potential problems" mentioned by Professor Preesip provide the opportunity to turn it into a research experiment.

Equipment

Pipetters or 10-mL graduated cylinders (for Step 1 dilutions)

50-mL volumetric flasks

Droppers

Squeeze bottles with distilled water

Q-tips™

10-mL graduated cylinders (preferably with ground glass stoppers)

Spectrophotometer (420 nm) and cuvettes

Chemicals

0.050 M solution of barium chloride

0.050 M solution of sodium sulfate

Experiment 18. Too Much Sodium in Soda Pop?

Author's Statement

This experiment supports the classroom discussion of solution preparation by dilution and the dilution calculations involved. It also reintroduces the topic of atomic spectroscopy, which may have been mentioned first in a discussion of electron configurations.

Important Note: If Experiment 19 is also on the students' schedule, it is important for the premise of that experiment that the Tasti Cola tested here NOT be high in sodium. This means that the soda pop sample should be taken from an ordinary soda pop can without spiking with additional sodium.

Equipment

25-mL volumetric flasks

Pipetters

Small beaker

Ultrasonic bath (or other means of degassing the soda pop)

Atomic absorption spectrophotometer

Graph paper (or a computer with graphing software)

Chemicals

1000 ppm sodium stock solution

Experiment 19. Conductivity, Odors, and Colors — Oh My!

Author's Statement

This experiment supports the classroom discussion of solution preparation and required calculations (including molar solutions by dilution or by weighing a pure solid, and percent solutions) as well as the concepts of solution conductivity and oxidation and reduction processes at electrode surfaces. There are 11 solutions to be prepared. Rather than have each student prepare all 11, it is suggested that students be allowed to work in groups with each group given a set of 2 to 4 solutions to prepare such that each group gains experience preparing molar solutions by both dilution as well as weighing a pure solid. The solutions are then shared by all groups so that each student gets the conductivity and accompanying observation experience for each.

Equipment

125-mL Erlenmeyer flasks with rubber stoppers

Top-loading balances

Conductivity apparatus, each consisting of the following components:

    a. Milliammeter (capable of measuring currents less than 30 mA as well as above 150 mA)

    b. 9-volt battery with snap-on adapter having wire leads (available at Radio Shack)

    c. Wire leads with alligator clamps on both ends (available at Radio Shack)

    d. 50-mL beaker or similar sized jar

    e. Two-hole rubber stopper to fit in the 50-mL beaker or jar

    f. Short graphite carbon rods positioned in the holes of the stopper (serving as electrodes) such that one end nearly touches the bottom of the beaker while the other protrudes through the top of the stopper when the stopper is positioned in the beaker

Chemicals

1.0 M acetic acid

Solid sodium hydroxide

2.0 M hydrochloric acid

Solid sodium chloride

0.5 M ammonium hydroxide

Solid potassium iodide

Solid copper sulfate pentahydrate

2.5 M sulfuric acid

Ethyl alcohol

Solid sodium bromide

Sugar

Experiment 20. Humid Fun for National Chemistry Week

Author's Statement

This experiment supports classroom discussions of chemical equilibrium and LeChatlier's Principle. Students also gain experience with a research activity in which things may go well or they may not. This should be a group activity in which a minimum number of solutions are prepared and only one set of test squares is created. However, each square, regardless of who prepared it, is checked by all students daily over a period of a week or longer. While there are 10 variations suggested, the number could easily be 20 or 30. Each student can prepare one square or a group of students can prepare a set of squares.

Equipment

Whatman #2 filter paper cut into 1-in. squares

Top-loading balances

Small beakers

Watch glasses

Fume hood

Electronic humidity indicator

10- or 20-gal fish tank

Test tubes

Hot water bath (250-mL beaker on hot plate)

Cold water bath (250-mL beaker filled with slushy ice-water mixture)

Chemicals

Cobalt chloride hexahydrate

Cobalt nitrate hexahydrate

Sodium chloride

Ammonium chloride

12 M HCl

Experiment 21. The Soil Project

Author's Statement

This experiment is meant to be a project covering several important concepts of laboratory measurement, including moisture by physical separation and weight loss, particle sizing, chemical testing, chemical separation and weight loss, pH, and residue analysis by weight loss. It is meant to be an exercise coinciding with the study of pH in the classroom. Since the moisture analysis requires a 24-h drying period, students must begin the project in advance of when most of the work is performed. The instructor should obtain and check soil samples ahead of time to ensure that there are some samples that contain carbonate. Samples may be obtained from the USDA National Soil Survey Laboratory in Lincoln, NE.

Equipment

Evaporating dishes	Squeeze bottle with distilled water
Labeling tape	Small glass vials (capable of holding 7 mL of solution)
Top-loading balance	10-mL graduated cylinders
Soil samples	50-mL Erlenmeyer flask (or other container
Drying oven	appropriate for shaker or stirrer to be used)
Tongs	Shaker or magnetic stirrer
Mortar and pestle	pH meters
2-mm particle sieve	Porcelain crucibles
Plastic sample bags	Hi-temp markers for labeling crucibles
Porcelain spot plates	Bunsen burners
125-mL Erlenmeyer flask	Analytical balances

Chemicals

　1 M HCl

　6 M HCl

Experiment 22. Identifying the Waste Acid

Author's Statement

This experiment is intended to support classroom discussions concerning acid-base neutralization and pH. Students should work in groups of two or three. Time may be an issue. To save time, the 0.050 M acid and base solutions can be prepared in advance. Also, each group can be assigned to perform one or two of the titrations instead of all three, and the data then shared.

Equipment

　125-mL Erlenmeyer flasks with stoppers (to contain the 0.050 M acid solutions)

　Pipetters or droppers suitable for measuring 1 mL of the concentrated acid solutions

　250-mL plastic bottles (or larger) for 0.050 M NaOH solutions

　50-mL burets

　pH meters with combination pH probes

　Materials for standardizing the pH meters (buffer solutions and small beakers)

　250-mL beakers

　Magnetic stirrers with stirring bars

　Ring stands with buret clamps

　25-mL graduated cylinders

　Graph paper (or microcomputers with graphing software)

Chemicals

　Unknown acid solution (either 5.0 M H_2SO_4 or 5.0 M H_3PO_4)

　0.050 M H_2SO_4 and 0.050 M H_3PO_4 (or higher concentrations to be diluted by students)

　0.10 M NaOH (or solid NaOH to be used to prepare a 0.10 M solution)

　Buffers solutions for standardizing pH meter

Experiment 23. The Production Advantage

Author's Statement

We are grateful to Dr. Karen Wosczyna-Birch of Tunxis Community College in Connecticut for this experiment. It is intended to support classroom discussions concerning reaction rates.

Equipment

　Baby bottles

　250-mL graduated cylinders

Top-loading balances

Buckets

Timers

Hot plates

Thermometers

Graph paper (or microcomputers with graphing software)

Chemicals

Three concentrations of hydrochloric acid (0.050 M, 0.10 M, 0.15 M)

Three forms of magnesium metal (ribbon, granules, powder)

Experiment 24. Does It or Doesn't It: The Fruit Juice Controversy

Author's Statement

We are grateful to Don Mumm of Southeast Community College, Lincoln, NE, for this experiment. It is intended to support classroom discussions of amino acids and general concepts of biochemistry, but also could be used as a demonstration of the separation of a mixture.

Equipment

Whatman #1 chromatography or filter paper, 25 × 16 cm

TLC plates (silica gel), 25 × 16 cm

Capillary tubes

Paper clips

Covered developing jar

Drying oven

Ruler

Chemicals

0.01 M solutions of leucine, methionine, valine, and lysine in 70% isopropyl alcohol

Fruit juice samples

Isopropyl alcohol — 70%

Ninhydrin solution (0.01% in butanol or isopropyl alcohol)

Experiment 25. The Fun Fat Analysis

Author's Statement

We are grateful to Don Mumm of Southeast Community College in Lincoln, NE, for this experiment. It is intended to support classroom discussions concerning fats and other general concepts of biochemistry.

Equipment

100 mL round bottom extraction flasks with stoppers

400-mL beakers (to hold extraction flask upright)

Top-loading balances

Weighing paper

Funnels

Hot water baths (hot plates with 600-mL beakers)

10-mL graduated cylinders

Freezer

Chemicals

HCl (concentrated)

Ethyl ether

Petroleum ether